General ENGLISH

for

GATE • PSUs

Also useful for

State public service commissions and other competitive examinations

MADE EASY PUBLICATIONS

MADE EASY Publications Pvt. Ltd.

Corporate Office: 44-A/4, Kalu Sarai (Near Hauz Khas Metro Station), New Delhi-110016
E-mail: infomep@madeeasy.in
Contact: 9021300500

Visit us at: www.madeeasypublications.org

General English for **GATE • PSUs**

First Edition: 2021
Second Edition: 2022
Third Edition: 2023
Fourth Edition: 2024

Contents

Section A | Grammar 01 - 155

Section B | Vocabulary 156 - 200

Section C | GATE Solved Questions 201 - 250

Section

A

GRAMMAR

CHAPTER 1 Tenses

Tense

Tense is a form of verb which tells us about the time of action.

1. The Present Tense
2. The Past Tense
3. The Future Tense

Aspects

1. Indefinite / Simple
2. Progressive / Continuous
3. Perfect
4. Perfect Progressive

> **NOTE**
> In fact there are only two tenses in English, but the future aspect is possible with the help of modal i.e. will / shall or ***Present Simple*** / ***Present Progressive***.

Simple Present introduced by

Always, Everyday, Sundays, Mondays, Often, Usually, Seldom, .

Formation

1. **Singular Subject :** ***He, She, It, Name.***
2. **Plural Subject :** ***You, we, They.***
3. ***'I'*** is a singular subject, but treated as plural in the present tense.

Rule : 1 **(Singular sub + V_1 + s/es + ob)**

V_1 [main verb/base form/principal verb/root verb] is always plural; we add s/es to make it singular.

1. ***Go*** : Plural verb
2. ***Goes*** : Singular verb

EXAMPLES —

1. Mohan ***goes*** to office.
2. She cooks food.
3. One of my friends lives in Delhi.

Rule : 2 **(Plural sub + V_1 + object)**

1. They go to temple.
2. We help the poor.

(Negative)

[To say no to something or somebody]

(Singular sub + does not + V_1 + ob)

EXAMPLES —

1. She does not go to temple.
2. I do not compose a song.

Never = Not any time in Present, Past or Future.

1. She never comes on time.
2. They never help the poor.

Interrogative / Interrogative negative [Asking Question]

Yes / No Type [Do/ Does/ Sub + V_1 + ob?]

1. Does she write a letter?
2. Do you play hockey?

WH type [WH + do / does + sub + V_1 + ob?]

Where do you live?

When does she go to office?

Why do you not go to school?

> **NOTE**
>
> ***Who*** functions as a singular sub. **(Who + V_1 + s/es + ob?)**

Who teaches you English?

Usage:

1. **Present Habits**
   - Cindy cooks food [Again and again]
2. **News paper headlines**
   - India wins by three wickets.
3. **Near future**
   - She goes to Holland next month.
4. **Universal Truth**
   - The sun rises in the east.

Present Continuous Tense

The present continuous tense is used to indicate an action which is going on at the time of speaking.

EXAMPLE —

She is writing a letter. *[Now]*

To understand it better, please read the following points carefully:

1. Action should be going on while speaking.
2. Action should be deliberate.
3. Action should be temporary.
4. Sometimes action may be incomplete while speaking.
5. Verb should be dynamic

EXAMPLE —

I am writing a novel. [It maybe the writer is not writing a novel while speaking]

Rule : **(Subject + is / am / are + V_1 + ing + object)**

EXAMPLES —

1. She is writing a letter.
2. I am listening to the music.
3. They are watering the plants.

VERB	
Action	**State**
Reading	Taste
Writing	Hate
Playing, etc.	Prefer, like, etc.

To indicate 'state' we do not use progressive tense.
Some verbs are not used in progressive tenses. They are called **Stative Verbs**.

Stative Verbs

like, dislike, know, belong to, love, hate, detest, realise, seem, appear, sound, agree, mind, believe, remember, forget, [look = seem] prefer, resemble, see, hear, smell, taste, astonish, have, recognise, own, wonder, understand, think.

☒ **Wrong** : I am understanding your problem.

☑ **Right** : I understand your problem.

☒ **Wrong** : He is wondering to see her shouting at the beggar.

☑ **Right** : He wonders to see her shouting at the beggar.

NOTE

Some ***Stative verbs*** can be used in continuous to indicate temporary action.

1. **Think (Stative) : Opinion / believe**

   Ex: I ***think*** you are right **[correct]**

   It ***means***, It is my opinion that you are right.

   Think (Dynamic) : Consider, or to have something on mind

   Ex: I am ***thinking*** about my friend.

2. **See (Stative) : Understand / see with your eyes**

   Ex: I ***see*** what you mean.

   I ***see*** a donkey on the road.

   See **(Dynamic) : To meet**

   I am ***seeing*** her tomorrow.

Present Perfect Tense

This tense is used to indicate an action which is finished just now / recently.

[Sub + has / have + V_3 + ob] **[V_3 = Past Participle]**

1. [He / She / It / Name + has]
2. [I, you, we, they + have]

EXAMPLES —

1. He has written a letter.
2. They have composed a song.

Negative [Subject + has / have + not + V_3 + Ob]

He has not written a letter.
They have not composed a song.

Interrogative / Negative
WH + (has / have + subject + V_3 + object?)

1. Have they watered all the plants?
2. Why have you not eaten food yet?

Errors in the use of present perfect.

Rule : 1 **(imperative sentence + when + subject + has / have + V_3 + object.)**

☒ **Wrong** : Take this medicine when you ate food.
☑ **Right** : Take this medicine when you have eaten food.

☒ **Wrong** : Ask him to go to school when he completes his homework.
☑ **Right** : Ask him to go to school when he has completed his homework.

Rule : 2 **(Gone / been)**

Gone : To go somewhere and be there.
Been : To go somewhere and come back.

☒ **Wrong** : Have you ever gone to Holland?
☑ **Right** : Have you ever been to Holland?
☒ **Wrong** : Cindy has been to office, so Rochester is waiting for her.
☑ **Right** : Cindy has gone to office, so Rochester is waiting for her.

Rule : 3 **(It is the first time, the second time + ... + Subject + has / have + V_3 + object.)**

☒ **Wrong** : It is the third time, I saw her melting in his arms.
☑ **Right** : It is the third time, I have seen her melting in his arms.

Rule : 4 **(This is the only time + Subject + has / have + V_3 + object.)**

☒ **Wrong** : This is the only time Cathy went to America to meet her Sunday husband.
☑ **Right** : This is the only time Cathy has gone to America to meet her Sunday husband.

Rule : 5 **(Present Perfect + since + subject + V_2 + object.)**

☒ **Wrong** : Five years have passed since I have seen him reading a detective novel.
☑ **Right** : Five years have passed since I saw him reading a detective novel.

Rule : 6 **(Subject + has / have + V_3 of stative verb + object + since / for + time.)**

☒ **Wrong** : I know her since childhood.

☒ **Wrong** : I have been knowing her since childhood.

☑ **Right** : I have known her since childhood.

☒ **Wrong** : I have this car since 2015.

☑ **Right** : I have had this car since 2015.

☒ **Wrong** : How long do you love Belinda?

☑ **Right** : How long have you loved Belinda?

Rule : 7 **(Do not use the Present Perfect Tense with the Past Time)**

☒ **Wrong** : Lydia has left her home town bag and baggage ten years ago.

☑ **Right** : Lydia left her home town bag and baggage ten years ago.

☒ **Wrong** : Meera has written a letter yesterday.

☑ **Right** : Meera wrote a letter yesterday.

Present Perfect Continuous

This tense is used to indicate an action which starts in the past and continues while speaking.

1. **[Subject + has / have + been + V_1 + ing + object + since / for + time]**
2. **Usage of [since / for]**

   'Since' is used to denote point of time.

   - Morning, evening, noon
   - Sunday, Monday, ...
   - January, February, ...
   - 2010, 2012, ...
   - 2 o'clock, 3 o'clock, ...
   - Birth, death, childhood, boyhood, ...
   - Festivals : Holi, Diwali, ...
   - Century : 19th century, 18th Century, ...
   - **O'clock** — of the clock [5 o'clock means 5 of the clock]

REMEMBER

Since + last	For + the last

For is used to denote period of time such as for two months, years, hours, etc.

NOTE

The Present Perfect continuous is not used with '***Stative Verbs***'.

☒ **Wrong** : They are watering the plants since morning.

☑ **Right** : They have been watering the plants since morning.

☒ **Wrong** : How long are you sitting in this classroom?

☑ **Right** : How long have you been sitting in this classroom?

EXAMPLES —

1. They have been teaching in this school for the last two years.
2. Mother has been cooking food since morning.

   Negative : They have not been reading this book for two days.

Interrogative / Negative

Has it been raining since morning?
Why have they not been reading their books since 5 o'clock?

Simple Past Tense

The simple past tense is applied to indicate
Past Habits
Past Events

Past Habits [Subject + used to + V_1 + object]

EXAMPLES —

'She used to smoke'. means now she does not smoke or we can say she was in the habit of smoking, but now, she does not smoke.
We used to read novels

Negative **[Subject + did not + use to + V_1 + object]**

- She did not use to smoke. *OR* • She used not to smoke.

Interrogative / Negative **[Did + subject + use to + V_1 + object]**

- Did she use to take tea?
- Did he not use to read books?

Past Events [Subject + V_2 + object]

EXAMPLES —

1. He wrote a number of books last year.
2. Ram went to Agra to see the Taj Mahal.
3. They helped the poor yesterday.

Negative

[Subject + did not + V_1 + object]
[Subject + never + V_2 + object]

- He did not go to Agra to see the Taj Mahal.
- She did not help me yesterday.
- I never saw him smoking.

Interrogative / Negative

[Did + subject + V_1 + object]

- Did he come here yesterday?
- Where did you go yesterday?
- Why did you not go there?

SPECIAL USAGE

It is time	
It is high time	**+ Subject + V_2 + Object.**
It is about time	

EXAMPLES —

☒ **Wrong** : It is time she goes to office.
☑ **Right** : It is time she went to office.

☒ **Wrong** : It is high time we help the poor.
☑ **Right** : It is high time we helped the poor.

Do not shift the tense without any proper reason.

EXAMPLE —

☒ **Wrong** : He went to office, completed his work, and report it to the manager.
☑ **Right** : He went to office, completed his work and ***reported*** it to the manager.

Past Continuous Tense

[When + subject + V_2 + object, sub + was / were + V_1 + ing + object.]

> **REMEMBER**
> 1. ***I, he, she, it, name + was***
> 2. ***You, we, they + were***

EXAMPLES —

☒ **Wrong** : When I saw him, he wrote an application to his principal.
☑ **Right** : When I saw him, he was writing an application to his principal.

☒ **Wrong** : When you rang me up, I sat in the classroom.
☑ **Right** : When you rang me up, I was sitting in the classroom.

OR

I was sitting in the classroom when you rang me up.

Past Perfect Tense

This tense is applied to indicate two actions finished in the past before or after.

Rule : 1 **(Subject + had + V_3 + object + before + subject + V2 + object)**

EXAMPLES —

☒ **Wrong** : He wrote a letter before she went to school.
☑ **Right** : He had written a letter before she went to school.

Negative

He had not written a letter before she went to school.

Interrogative / Negative

Had he written a letter before she went to school?
Had he not written a letter before she went to school.

Rule : 2 **(Subject + V_2 + object *after* Subject + had V_3 + object)**

EXAMPLES —

☒ **Wrong** : Rochester composed a song after mother went to market.
☑ **Right** : Rochester composed a song after mother had gone to market.

Negative

She did not cook food after mother had gone to office.

Interrogative / Negative

- Did she cook food after mother had gone to office?
- Why did he not go to market after they had completed their homework?

Rule : 3 **It was the first time, the second time + (Subject + had + V_3 + object.)**

EXAMPLE —

☒ **Wrong :** It was the third time, he went to Agra to see the Taj Mahal.

☑ **Right :** It was the third time, he had gone to Agra to see the Taj Mahal.

Past Perfect Continuous Tense

This tense is used to indicate an action that began before a certain point in the past and continued up to that time.

[Subject + had been + V_1 + ing + object + since / for + time.]

EXAMPLES —

1. It had been raining for two hours.
2. They had been teaching in this college for the last ten years.
3. He had been reading this novel since morning.

Negative

- It had not been raining for two hours.

Interrogative / Negative

- Had she not been reading since morning?
- Had it been raining for two hours?

☒ **Wrong :** They were writing an essay since 5 o'clock.

☑ **Right :** They had been writing an essay since 5 o'clock.

☑ **Right :** Sheela had been teaching in this college since last ten years.

Simple Future Tense

Simple Future Tense is used to talk about an action that may take place in the coming time (i.e. future). There are several ways to express it:

1. **Present indefinite + coming time**

   Example —

   Cindy goes home next month.

2. **Present continuous + coming time.**

   Example —

   Sandra is coming here tomorrow.

3. **About to :**

   Example —

   The train is ***about to*** arrive.

   ☒ **Wrong :** The train is ***about to*** arrive in ten minutes.

   ☑ **Right :** The train arrives in ten minutes.

   OR

   The train is arriving in ten minutes.

4. **Going to :**

   Example —

   - The match is ***going to*** start.
   - I am ***going to*** tell you a story.

5. **Intend (verb)**

   Example —

   - I ***intend*** to buy a new car.

6. Will / Shall

EXAMPLES —

- I *will* go there.
- We *will* help them.

Negative *[subject + will not + V_1 + object]*

- We will not help them.
- She will not go to office tomorrow.

Interrogative / Negative *[will + subject + V_1 + object]*

- *Will* she come here tomorrow? *[shall + I / we + V_1 + object]*
- *Shall I* use your pen?

  OR
- *May I* use your pen?

Future Continuous Tense

This tense is used to indicate an action which is supposed to be in progress in the coming time.

[Subject + will + be V_1 + ing + object]

EXAMPLES —

- He *will be* reading a book.
- They *will be* composing a song.

Negative

- He *will not be* reading a book.

Interrogative / Negative

- *Will he be* reading a book?
- *Will he not* reading a book?

Future Perfect Tense

The future perfect tense indicates an action that will have been completed at some point in the future.

[Subject + will have + V_3 + object]

EXAMPLES —

- The parade *will have* ended by the time Mohan gets out of bed.
- He *will have* watered all the plants before Mohan comes.

☒ **Wrong** : Sandra *will have* composed a poem before mother cook food.
☑ **Right** : Sandra *will have* composed a poem before mother cooks food.

Future Perfect Continuous Tense

We use the future perfect continuous tense to show that something will continue until a particular event or time in the future.

[Subject + will have been + V_1 + ing + object + by the / for + time]

☒ **Wrong** : It ***will have been*** raining since morning.

☑ **Right** : It ***will have been*** raining by the morning.

☑ **Right** : Alison ***will have been*** teaching in this school for five years.

NOTE

Do not use since with the ***Future Perfect Continuous***.

Negative

- He ***will not have been*** reading this book by the noon.

Interrogative / Negative

- ***Will he have been*** reading this book by the noon?
- ***Will he not have been*** reading this book by the noon?

PRACTICE EXERCISE Tenses

Instructions : Out of the following four sentences, select the most suitable sentence with respect to grammar and usage.

1. (a) He has left his home town with bag and baggage last year.
(b) He has leave his home town with bag and baggage last year.
(c) He had left his home town with bag and baggage last year.
(d) He left his home town with bag and baggage last year.

2. (a) How long do you know him?
(b) How long have been knowing him?
(c) How long have you known him?
(d) How long are you knowing him?

3. (a) I am understanding your problem.
(b) I understand your problem.
(c) I understanding your problem.
(d) I understands your problem.

4. (a) I am reading this book since noon.
(b) I do read this book since noon.
(c) I had read this book since noon.
(d) I have been reading this book since noon.

5. (a) They have been teaching in this school since the last ten years.
(b) They have been teaching in this school for the last ten years.
(c) They have been teaching in this school for the last ten year.
(d) They have being teaching in this school since the last ten years.

6. (a) Return this book to me when you read it.
(b) Return this book to me when have you read it.
(c) Return this book to me when you have read it.
(d) Return this book to me when you will read it.

7. (a) It is the only time she goes to office.
(b) It is the only time she went to office.
(c) It is the only time she has gone to office.
(d) It is the only time she is going to office.

8. (a) It is high time he writes a letter.
(b) It is high time he wrote a letter.
(c) It is high time he was writing a letter
(d) It is high time he is writing a letter.

9. (a) It will have been raining since morning.
(b) It will have been raining for morning.
(c) It will have been raining by the morning.
(d) It will has been raining by the morning.

10. (a) Sandra cooked food before you came.
(b) Sandra has cooked food before you came.
(c) Sandra had cooked food before you come.
(d) Sandra has cooked food before you comes.

Answer Key ➤ Tenses

1. (d) 2. (c) 3. (b) 4. (d) 5. (b) 6. (c)
7. (c) 8. (b) 9. (c) 10. (c)

Explanations ➤ Tenses

1. (d)

Present perfect is not followed by past time, so options (a) and (b) are wrong. For single action we don't use past perfect tense, so option (c) is wrong.

For past time we use past indefinite tense so option (d) is the best option choice.

2. (c)

'How long' is followed by either perfect tense or perfect continuous tense. So options (a), (b) and (d) are wrong.

According to rule option (c) is the best option choice.

3. (b)

Understand is a stative verb so it has no continuous tense, so options (a) and (c) are wrong.

'*I*' takes plural verb (V_1) so option (d) is wrong. Option (b) is the best option choice.

4. (d)

In case of since or for, perfect tense is required, so options (a), (b) and (c) are wrong.

According to rule option **(d)** is the best option choice.

5. (b)

The Rule says :

For + The last
Since + last

According to the above rule

Option (a) is wrong.

Option (c) is wrong because it contains ten year instead of the ten years.

Being is not used with perfect continuous tense, so option (d) is wrong.

Thus according to rule option (b) is the best option choice.

6. (c)

The Rule says :

[Imperative sentence + When + Present Perfect Tense]

According to above rule options (a), (b) and (d) are wrong.

Option (c) is the best option choice.

7. (c)

The Rule says :

[It is the only time + Present Perfect Tense]

Options (a) and (b) are wrong.

Option (c) is the best option choice.

8. (b)

The Rule says :

[It is high time + Past Indefinite Tense]

According to above rule options (a), (c) and (d) are wrong.

Option (b) is the best option choice.

9. (c)

We do not use since with future perfect continuous tense, so option (a) is wrong.

We do not use for before morning. So option (b) is wrong.

Will is not followed by has, so option (d) is wrong.

We use by the at the place of since with future perfect continuous tense, so option (c) is the best option choice.

10. (c)

The Rule says :

[Subject + had + V_3 + object + before + subject + V_2 + Object]

According to rule of past perfect tense

Options (a), (b) and are wrong.

Option (c) is the best option choice.

■

CHAPTER 2

Articles

Articles

In fact like quantifiers, articles belong to the wider class of determiners. There are two articles in English.

Indefinite Article	Definite Article
a / an	*The*

Indefinite Article a / an

We use *a / an* before a singular countable noun when it is not specified, or when we mean no matter which one.

Some Important Key points

■ A **Do not use *a / an*** before an uncountable noun.

Such as : Information, soap, chalk, bread, etc.

☒ **Wrong** : *an* information, *a* soap, *a* chalk, *a* bread.

☑ **Right** : *a* piece of information
a bar of soap
a stick of chalk
a loaf of bread

Do not use *a / an* before an adjective which is without a **singular countable noun**.

☒ **Wrong** : He is a unique.

☑ **Right** : He is unique.

OR He is a unique person.

Do not use *a / an* before a **plural countable noun**.

Such as : Oxen, trousers, theses, oases, teeth, feet, etc.

☒ **Wrong** : He wrote a theses.

☑ **Right** : He wrote theses.

OR He wrote a thesis.

☒ **Wrong** : I saw an oxen.

☑ **Right** : I saw several oxen.

OR I saw an ox.

Do not use *a / an* after the following **phrases**.

kind of, sort of, post of,a number of, the number of	+ *no article* + *Noun*

☒ **Wrong** : Shalini has applied for the post of an editor.
☑ **Right** : Shalini has applied for the post of editor.

☒ **Wrong** : The number of the Crimes seems to be increasing by leaps and bounds in metropolitans.
☑ **Right** : The number of crimes seems to be increasing by leaps and bounds in metropolitans.

■ B Read the following ***fixed phrases*** with or without articles.

In the morning	**On foot**
In the evening	**By bus, buy car, by plane**
At noon	**Make a noise**
At night	**In difficulty**
At midnight	**In trouble**
Tell a lie	**In a hurry**
Speak the truth	**On horse back**

■ C We use ***a/ an*** before a **singular countable noun** when it is mentioned for the first time,

Ex. :
- Once there was a king.
- I bought a car.

■ D ***A / an*** can be used before ***food, lunch, breakfast, dinner*** if they are preceded by an adjective.

Ex. :
- He gave me a delicious food.

Rules Regarding a / an

- 'A' is used before a singular countable noun which begins with a consonant sound.
- 'An' is used before a singular countable noun which begins with a vowel sound.

Examples —

an S.D.M	a European citizen
an M.P.	an L.L.B student
an hour	a tube of toothpaste
an heir	a sheet of paper
an Umbrella	an English woman
a Unique woman	

few little	a few a little	the few the little
Almost nothing	some	Al that has been defined
few ; ***a few*****,** ***the few*** + Plural countable noun		**little,** ***a little*****,** ***the little*** + Uncountable noun

☒ **Wrong** : I was very thirsty, so she gave me little water to drink.

☑ **Right** : I was very thirsty, so she gave me a little water to drink.

☒ **Wrong** : I gave him a five hundred rupee note so that he might purchase few books.

☑ **Right** : I gave him a five hundred rupee note so that he might purchase a few books.

☑ **Right** : The few books that you gave me were very interesting.

Definite Article : The

The	before **consonant sound** (ð)
	before **vowel sound** (ði)

'The' can be used before

1. Uncountable noun
2. Singular countable noun
3. Plural countable noun

'The' is used before a noun / nouns to make it / them specified.

- We use ***a / an*** when a noun is mentioned for the first time and thereafter we are *'The'*.

EXAMPLE —

- I bought a new car last month after I got the car home, one of its types went flat.

Some important rules regarding 'The'

Rule : 1 *'The'* is used before the name of mountain ranges.

Ex. : The Himalayas, The Alps, The Vindhyas, etc.

☒ **Wrong** : The mount Everest is the highest peak of ***the*** Himalayas.

☑ **Right** : Mount Everest is the highest peak of ***the*** Himalayas.

Rule : 2 *'The'* is used before proper noun such as Ram, Mohan, Agra, Delhi, etc. to make a comparison.

Example –

- Kalidas is the Shakespeare of India.
- Kashmir is the Switzerland of India.

Rule : 3 *'The'* is used adverbially with comparative adjective in such phrases.

Example –

- The sooner, the better.
- The higher you go, the cooler you feel.

Rule : 4 *'The'* is used before a singular countable noun to represent the whole class.

EXAMPLES —

- The ant is industrious.
- The camel can store water.

The ant = All the ants.

The camel = All the camels.

Rule : 5 We do not use *'The'* before gerund, but we use *'The'* before gerund when it is followed by *'of'* preposition.

EXAMPLES —

- Reading is a good habit.
- Smoking is injurious to health.

☒ **Wrong** : The swimming is a useful exercise.
☑ **Right** : Swimming is a useful exercise.

☒ **Wrong** : Reading of this novel will help you understand the real condition of the women in the 18th century.
☑ **Right** : The reading of this novel will help you understand the real condition of women in the 18th century.

Rule : 6 We use *'The'* before a comparative degree in case of a choice.

[The + comparative degree + of the two]

☒ **Wrong** : She is wiser of the two sisters.
☑ **Right** : She is the wiser of the two sisters.

☒ **Wrong** : This book is better of the two.
☑ **Right** : This book is the better of the two.

Rule : 7 *'The'* is applied before the names of the ***religious books*** if they are not preceded by their writers' names.

EXAMPLE —

The Gita, The Mahabharat, The Bible, The Quran, etc.

Don't say - The Ramayana. *Say :* Balmiki's Ramayana.

Rule : 8 When we talk about a single person who posses two post, the article *'The'* will be used once at the initial position.

EXAMPLE —

- The principal and class teacher is not coming today.

NOTE **Here the principal and class teacher is the same person.**

Rule : 9 If we talk about two people the article *'The'* will be used twice.

EXAMPLE —

- The principal and the class teacher were absented themselves from the school yesterday.

Rule : 10 When a person posses the quality of someone *'The'* is used

EXAMPLES —

- The mother in her.
- The Judge in him.
- The father in him.

Rule : 11 **The** is used before the name of

Rivers : The Ganga, The Yamuna, The Kaveri, etc.

Seas : The Red sea, The Black sea, The Caspian, etc.

Oceans : The Indian ocean, The Pacific, The Atlantic, etc.

Gulfs : The Gulf of Mexico, The Persian Gulf, etc.

Deserts : The Sahara, The Kalahari, etc.

Musical instruments : The Guitar, The Sitar, The Mouthorgan, etc.

Historical buildings : The Qutub Minar, The Redfort, etc.

Political groups : The B.S.P., The AAP, The Congress, The CPI, etc.

Religious groups : The Sikhs, The Muslims, The Hindus, etc.

News papers : The Times, The Times of India, The Indian Express, etc.

Magazines : The Times, The Week, etc.

Inventions : The telephone, The computer, The wireless, etc.

Express trains : The Rajdhani express, The Garibrath, The Kalka express, etc.

Rule : 12 **The** is used before an adjective when it represents a class of people.

EXAMPLES —

- The rich = All the rich people.
- The poor = All the poor people.
- The blind = All the blind people.
- The mad = All the mad people.
- The deaf = All the deaf people

Zero Article - No Article

By *'zero article'* we mean no article is required.

1. **No article is used** before the names of languages.

   English, French, Spanish, Pali, Malayalam

   But, ***The*** is used before the names of languages that end in, ***N***, ***CH***, ***SH*** to indicate the people of that country.

   - The English = English people
   - The French = People of France
   - The Spanish = People of Spain

   EXAMPLE —

   English is the language of the English.

2. **No article is used** before school, church, temple, mosque, prison, bed, hospital, market, when they are used for their primary purposes.
   - He goes to school (To study)
   - He goes to the school (For some other purpose)

☒ **Wrong** : He met with an accident, so he was admitted to the hospital.
☑ **Right** : He met with an accident, so he was admitted to hospital.

☒ **Wrong** : Meera went to school to meet her son.
☑ **Right** : Meera went to the school to meet her son.

3. **No article is used** before proper nouns : Ram, Mohan , Delhi, etc.
4. **No article is used** before material nouns : Iron, gold, silver, etc.
   - ***But we can say :*** The gold of your ring seems to be of good quantity.
5. **No article is used** before abstract nouns : Honesty, beauty, kindness, etc.
   - ***But we can say :*** The beauty of Kashmir attracts everyone.
6. **No article is used** before the names of States / Countries : India, England, Bihar, Haryana, etc.
   - ***But remember the followings :*** The Punjab, The U.K. The USA, The Hague, The Netherlands, The West Indies.
7. **No article is used** before the names of diseases : Fever, Cancer, etc.
   - ***But we can say:*** The Mumps, The Rickets, The Flu, etc.
   - ***Before noun complement such as :*** Appoint, make, elect, select, etc.

☒ **Wrong** : We made him a chairman.
☑ **Right** : We made him chairman.
☑ **Right** : They elected him president.

8. **No article is used** before the names of seasons : Summer, Winter, Spring, Fall, etc.
   - ***But we can say:*** Rita came crying in the fall of 1999.

☒ **Wrong** : Now that she is living in her own flat, she cleans the windows, twice a week in the summer and once a week in the winter.
☑ **Right** : Now that she is living in her own flat she cleans the windows twice a week in summer and once a week in winter.

9. **No article is used** before the names of university : J.N.U., CCS University, M.D.U., Oxford, Bristol, etc.
   - ***But we can say:*** The University of Oxford.
     The university of Delhi.

NOTE 1 **Deliberate omission of articles**

The articles are left out whenever you want to save space, time or money.

Such as in : Headings, Telegrams, Notes and labels etc.

Example –

- Police question youth after Delhi museum explosion.
- Ring bell and wait.

NOTE 2 **We do not use article before 'Man' when we mean by it man or woman (wider sense).**

We can say :

Example –

1. Man is a social animal.
2. Man is mortal.

☒ **Wrong** : It is a well known fact that a man is the only animal on this earth that can speak and think about anything.

☑ **Right** : It is a well known fact that man is the only animal on this earth that can speak and think about anything.

PRACTICE EXERCISE Articles

Instructions : Out of the following four sentences, select the most suitable sentence with respect to grammar and usage.

1. (a) I gave him a five hundred rupee note so that he might purchase a few books.
(b) I gave him a five hundred rupee note so that he might purchase few books.
(c) I gave him a five hundred rupee note so that he might purchase few book.
(d) I gave him a five hundred rupee note so that he might purchase a few book.

2. (a) Little knowledge that he got from newspapers and magazines was not sufficient for him even to get pass marks.
(b) A little knowledge that he got from newspapers and magazines was not sufficient for him even to get pass marks.
(c) The little knowledge that he got from newspapers and magazines was not sufficient for him even to get pass marks.
(d) Little knowledge which he got from newspapers and magazines was not sufficient for him even to get pass marks.

3. (a) He is a honest.
(b) He is an honest.
(c) He is a honest man.
(d) He is honest.

4. (a) She bought a toothpaste.
(b) She bought a tooth pastes.
(c) She bought a tubes of toothpaste.
(d) She bought a tube of toothpaste.

5. (a) Everybody knows that a man is a social animal.
(b) Everybody knows that an man is a social animal.
(c) Everybody knows that man is social animal.
(d) Everybody knows that the man is a social animal.

6. (a) She came crying in the fall of 1999.
(b) She came crying in fall of 1999.
(c) She came crying in fall of the 1999.
(d) She came crying in a fall of 1999.

7. (a) She is wiser of the two sisters.
(b) She is the wiser of the two sister.
(c) She is the wiser of the two sisters.
(d) She is a wiser of the two sisters.

8. (a) He comes here by the train.
(b) He comes here by a train.
(c) He comes here by an train.
(d) He comes here by train.

9. (a) He went to school to meet his son.
(b) He went to a school to meet his son.
(c) He went to the school to meet his son.
(d) He went to an school to meet his son.

10. (a) She has applied for the post of an editor.
(b) She has applied for the post of a editor.
(c) She has applied for the post of the editor.
(d) She has applied for the post of editor.

Answer Key ➤ Articles

1. (a)	2. (c)	3. (d)	4. (d)	5. (c)	6. (a)
7. (c)	8. (d)	9. (c)	10. (d)		

Explanations ➤ Articles

1. **(a)**

The correct phrase is ***a few books***, so option (b) is wrong.

'Few book' is also wrong phrase so option (c) is wrong.

'A few book' is also a wrong phrase so option (d) is wrong.

Thus, option (a) is the best option choice.

2. **(c)**

The correct phrase here is ***'The little knowledge'***, so options (a), (b) and (d) are not correct.

Thus, option (c) is the best option choice.

3. **(d)**

'Honest' is an adjective, so we cannot use ***'an'*** before it, if thus option (a) and (b) are wrong.

Option (c) is wrong because we cannot use ***'a'*** before ***'honest man'***.

Thus, option (d) is the best option choice.

4. **(d)**

'Toothpaste' is an uncountable noun, so option (a) and (b) are wrong.

We cannot say a tubes of toothpaste, so option (c) is wrong.

Thus, option (d) is the best option choice.

5. **(c)**

When ***'Man'*** is used in wider sense no article is required.

Man is a social animal is a correct sentence, so option (a), (b) and (d) are wrong.

Thus, option (c) is the best option choice.

6. **(a)**

The correct phrase is ***'in the fall of 1999'***, so option (b), (c) and (d) are wrong.

Thus, option (a) is the best option choice.

7. **(c)**

The Rule says :

The + Comparative degree of an adjective + of the two.

According to the above mentioned rule options (a), (b) and (d) are wrong.

Thus, (c) is the best option choice.

8. **(d)**

The correct phrase is ***by train***, so option (a), (b) and (c) are wrong.

Thus, option (d) is the best option choice.

9. **(c)**

We apply article ***'the'*** before school, when it is used as a building. So option (a) is wrong.

'a school' is also a wrong phrase here, so option (b) is also wrong.

'an school' is also a wrong phrase as ***'an'*** cannot come before school, so option (d) is wrong.

Thus, option (c) is the best option.

10. **(d)**

No article is to be used after post of so option (a), (b) and (c) are wrong.

Thus, option (d) is the best option choice.

CHAPTER 3

Nouns

Nouns

Nouns are naming words, they can name anything that we can see, touch, and feel.

Such as

Mohan, Sohan, Agra, Delhi, Honesty, Beauty, Childhood, etc are nouns.

Types of Nouns

1. **Proper Noun :** A Proper noun is the name of a particular person or thing, such as ***Meera, Deepak, Agra, Meerut*** etc.
   Proper noun begins with capital letter.
2. **Common Noun :** A common noun is a name shared by every person or thing of the same class.
   Such as ***man, woman, dog, river, table, pen etc.***
3. **Collective Noun :** Collective noun is (also called group noun) the name of a group of people.
   Such as ***class, team, army, committee, mob, team, crowd, etc.***
4. **Abstract Noun :** Abstract noun is the name of quality, state or idea.
   Such as ***beauty, wisdom, childhood, kindness, etc.***
5. **Material Noun :** Material noun refers to a material or substance from which things are made such as
   Such as ***gold, iron, silver, copper, plastic, etc.***

Genders

1. **Feminine gender :** Indicates ***female sex***
   Such as ***woman, girl, bitch, hen, cow, etc.***
2. **Masculine gender :** Indicates ***male sex***
   Such as ***man, boy, dog, bull etc.***
3. **Neuter gender :** Indicates an object that is ***neither male nor female,***
   Such as ***pen, book, table.***
   Birds, wild animals, baby (whose sex is not known) are also included in neuter gender.
4. **Common gender :** Indicate either man or woman
   Such as ***baby, student, nurse, homemaker, poet, actor, doctor, parent, etc.***
   poet, actor, homemaker, nurse etc. are included in common gender
   EXAMPLE –
   - He is an actor ☑
   - She is an actor ☑

Points to Remember

Some nouns are personified (They are non living but considered living)

Male category : Death, sun, winter, summer, time.

Female category : Moon ship, yacht, spring, nature, earth, name of the country except France.

EXAMPLES –

1. ***Death*** has put ***his*** icy hands on her rosy lips.
2. The ***ship*** has lost ***her*** way in the tempest.

Remember

- Some nouns are countable and some are uncountable, (that cannot be counted eg. milk, water, nouns are always singular.)
- Countable nouns (That can be counted such as one pen, three books, etc.) are singular and plural.
- Some countable nouns have their fixed plural forms.

Cases of Nouns

The case of a noun is form of a noun which shows its relationship to the other word in a sentence.

A noun has the following cases :

1. Nominative case
2. Objective case
3. Dative case
4. Possessive case

1. Nominative case :

When a noun is used as the subject of a verb it is called subjective case or nominative case.

Ram goes to school.

↓

Subjective case / Nominative case

2. Objective case :

When a noun is used after the verb it is said to be in objective case. Nominative case and objective case of nouns are same

Ram teaches me English.

↓

Nominative case

He helps Ram.

↓

Objective case

3. Possessive case :

Possessive case is used to show ownership.

- ***Ram's book*** is on the table.
- ***The legs of this table*** are broken.

4. Dative case :

When a noun indicates the indirect object of the verb, it is said to be in the dative case.

Sandra gave **Candida** a bouquet of flowers.

↓ Indirect Object

Numbers

There are two numbers in English.

1. Singular number
2. Plural number

1. Singular number :

A noun that denotes only one person or thing is said to be in the singular number. Such as ***man, pen, girl, toy, chair, etc.***

2. Plural number :

A noun that denotes 'more than one' person or thing is said to be in the 'plural number.' Such as ***men, pens, girls, toys, chairs, etc.***

Formation of Plural Nouns

1. **Some nouns have their fixed plural.**

Examples –

Singular Noun	Plural noun	Singular Noun	Plural noun
man	*men*	*mouse*	*mice*
woman	*women*	*mouse (computer)*	*mouses*
child	*children*	*madam*	*ladies*
ox	*oxen/oxes*	*person*	*people*
tooth	*teeth*	*crisis*	*crises*
foot	*feet*	*oasis*	*oases*
goose	*geese*	*thesis*	*theses*
louse	*lice*	*hypothesis*	*hypotheses*

2. **If a singular countable noun ends with** *SS, CH, X, O, Z, S* add *'es'* otherwise *'s'*.

Examples –

Singular Noun	Plural noun	Singular Noun	Plural noun	Singular Noun	Plural noun
bus	*buses*	*book*	*books*	*solo*	*solos*
fuss	*fusses*	*lip*	*lips*	*memo*	*memos*
box	*boxes*	*bed*	*beds*	*photo*	*photos*
watch	*watches*	*home*	*homes*	*kilo*	*kilos*

When *'ch'* is pronounced *'k'* and *s*.

monarch — monarchs stomach — stomachs arch — archs

3. **If a singular countable noun** ends with *'y'* add *'s'* if a vowel comes before y.
Examples –

Singular Noun	Plural noun
baby	*babies*
quality	*qualities*
country	*countries*
party	*parties*

days boys keys

Change *'y'* into *ies* if a consonant comes before *y*.

4. **If the noun ends with** *f/fe*, replace *f/fe* by *ves*.
Examples –

Singular Noun	Plural noun	Singular Noun	Plural noun
knife	*knives*	*belief*	*beliefs*
life	*lives*	*chief*	*chiefs*
loaf	*loaves*	*cliff*	*cliffs*
wife	*wives*	*roof*	*roofs*
		hoof	*hoofs*

Plural of Compound Noun

Two or more nouns are combined to form a compound noun.

Examples – Father in law, brother in law, passer by, boy friends, girl fiend etc.

1. **To make a compound noun plural**. Change the main noun into its plural form.

Singular Noun	Plural noun
father in law	*fathers in law*
brother in law	*brothers in law*
passer by	*passers by*
boy friend	*boy friends*

2. **Nouns ending in** *'ful'* add *'s'* to the end.

Singular Noun	Plural noun
Cupful	*Cupfuls*
Spoonful	*Spoonfuls*
Basinful	*Basinfuls*

3. **If the first element is man or woman, both elements are made plural.**

Singular Noun	Plural noun
Woman doctor	*Women doctors*
Woman driver	*Women drivers*
Man servant	*Men servants*

But we say, Man haters, woman killers etc.

4. **Foreign nouns and their plurals**

Singular Noun	Plural noun
Fungus	*Fungi*
Index	*Indexes / Indices (sign in Mathematics)*
Datum	*Data*
Medium	*Media*
Phenomenon	*Phenomena*
Agendum	*Agenda*
Memorandum	*Memoranda*
Stratum	*Strata*
Bacterium	*Bacteria*

Datum is not used in modern English.

5. **Wages is both singular and plural.**

Wages : *Singular* (result) *Plural* (labour charge)

Examples –

1. The wages of sin is death.
2. Their wage have been increased.

Formation of possessive nouns / Genitives

's / , / ' / of

Infact possessive nouns work as adjectives. They also show ownership. We use *'s/,* for living and personified nouns.

Examples –

1. Mohan's car is red.
2. Sohan's school seems to be tidy.

> Sohan's school, means, the school where Sohan is a student.

3. Nature's law.
4. Death's bed.

- Add just an apostrophe to the end of nouns that already end in 's' to make their possessive.

  Examples –
  - Boys' hostel
  - Girls' college.
  - Horses' stall
  - Kalidas' Shakuntalam etc.

- When two or more nouns share ownership, make sure to add the apostrophe *+'s'* to the last noun only.

  Example –

  ☒ **Wrong** : Ram's and Mohan's mother goes to temple.
  ☑ **Right** : Ram and Mohan's mother goes to temple.

- When two or more nouns show separate ownership. Add apostrophe ***+'s'*** to each noun.

  Example –

  ☒ **Wrong** : Ram and Mohan's wives are outstanding.
  ☑ **Right** : Ram's and Mohan's wives are outstanding.

- Add apostrophe ***+'s'*** at the end of hyphenated nouns and compound nouns.

  Examples –

  1. My mother-in-law's behaviour is very good.
  2. She stood before five attorneys general's office and knew her time was up.

- An apostrophe ***+'s'*** can be added to ***somebody, nobody, everybody*** etc, but if they end with else add the apostrophe ***+'s'*** after else.

  Examples –

  1. Everybody's concern is no one's concern.
  2. This is not my book it is somebody else's.

- We use ***(of)*** for non living things.

  Examples –

  1. Ram's leg. ☑
  2. Tables's leg ☒
  3. The legs of table. ☑
  4. The pages of this book are not so good. ☑
  5. Mohan found the bullet's exit point. ☒
  6. Mohan found the exit point of the bullet. ☑

- Double possessives:

  In English grammar the double possessive or double genitive is a phrase in which possession is indicated by the preposition ***'of'*** followed by the possessive form of a noun.

  Example –

  ☒ **Wrong** : Sheela is a friend of Mohan.
  ☑ **Right** : Sheela is a friend of Mohan's

- In time expressions, we use apostrophe ***+'s'***
  - One weeks's pay
  - Today's news
  - A month's holiday

- We can use apostrophe ***+'s'*** alone, when the missing noun stands for : ***store, shop, studio, office, restaurant, church*** etc.

  Please do go and buy a loaf at the baker's.

  Here ***'baker's'*** means ***baker's shop***.

- We do not add apostrophe ***+'s'*** with the adjectives when used as noun.

  ☒ **Wrong** : He intends to improve the poor's condition.
  ☑ **Right** : He intends to improve the condition of the poor.

Rules Regarding Nouns

Rule : 1 One third, two thirds, one fourth, three fourths *[Notice the difference]*

☒ **Wrong** : I have read three fourth of this novel.
☑ **Right** : I have read three fourths of this novel.

Rule : 2 If a noun precedes and follows a preposition, the noun should be singular.

☒ **Wrong** : Years after years rolled by, but she could not become rich.
☑ **Right** : Year after year rolled by, but she could not become rich.

Also remember : *day by day, door to door, word for word, hand to hand etc.*

Rule : 3 Always use singular countable noun after *'a / an'*

☒ **Wrong** : Candida is going on a ten days tour to Holland.
☑ **Right** : Candida is going on a ten day tour to Holland.

☒ **Wrong** : Christy went to the woods where he saw a forty feet tree.
☑ **Right** : Christy went to the woods where he saw a forty foot tree.

Rule : 4 Some nouns are always singular.

scenery, machinery, information, luggage, baggage, furniture, poetry, grass, soap, snow, toothpaste, chalk, etc.

☒ **Wrong** : I wanted to buy some soaps yesterday.
☑ **Right** : I wanted to buy some bars of soap yesterday.

☒ **Wrong** : I love reading poetries.
☑ **Right** : I love reading poetry / poems.

☒ **Wrong** : Cindy will provide you so many informations reading this case.
☑ **Right** : Cindy will provide you so many pieces of information regarding this case.

Rule : 5 Some nouns look singular, but they are plural.

Cattle, gentry, clergy, poultry, vermin, infantry, peasantry, police, cavalry, etc.

☒ **Wrong** : The gentry of the town was invited to the party by the president.
☑ **Right** : The gentry of the town were invited to the party by the president.

☒ **Wrong** : The police is running after the thieves.
☑ **Right** : The police are running after the thieves.

Rule : 6 Some nouns are same in their singular and plural forms.

sheep, swine, deer, fish, cod, reindeer, bison, moose, greenfly, salmon, trout, aircraft, shrimp, dice, offspring, rendezvous, series, etc.

☒ **Wrong** : I saw so many deers in the forest.
☑ **Right** : I saw so many deer in the forest.

☒ **Wrong** : A number of sheeps were grazing in the field.
☑ **Right** : A number of sheep were grazing in the field.

Rule : 7 Nouns in pair are considered plural in English.

Jeans, shorts, scissors, pants, trousers, binoculars, scales, etc.

☒ **Wrong** : Your trousers does not match your shoes.
☑ **Right** : Your trousers do not match your shoes.

☒ **Wrong** : Your scissor are lying on the table.
☑ **Right** : Your scissors are lying on the table.

Rule : 8 A group noun or collective noun can be singular or plural. The choice depends on whether we see the group as a whole or as individual people.

☒ **Wrong** : The committee have given its decision.
☑ **Right** : The committee has given its decision.

☒ **Wrong** : The family is delighted with their presents.
☑ **Right** : The family are delighted with their presents.

Rule : 9 ***One of + Plural noun + singular verb.***

☒ **Wrong** : One of my friend often goes to church on Sundays.
☑ **Right** : One of my friends often goes to church on Sundays.

☒ **Wrong** : Satyajit Ray, who conceived, co-authored, and directed a number of good films, was one of India's most talented film maker.

☑ **Right** : Satyajit Ray, who conceived, co authored, and directed a number of good films, was one of India's most talented film makers.

Rule : 10 ***The number of + plural noun + singular verb.***

☒ **Wrong** : The number of students are increasing rapidly.
☑ **Right** : The number of students is increasing rapidly.

Rule : 11 ***A number of + plural noun + plural verb.***

☒ **Wrong** : A number of crime are increasing in big cities.
☑ **Right** : A number of crimes are increasing in big cities.

Rule : 12 **(i)** If there is a ***mathematical*** number before ***dozen, hundred, thousand, million*** ..... ***do not make them plural.***

☒ **Wrong** : He bought ten dozens notebooks.
☑ **Right** : He bought ten dozen notebooks.

(ii) If there is no mathematical number before them, use ***[s + of]***

☒ **Wrong** : Dozens people were killed in that accident.
☑ **Right** : Dozens of people were killed in that accident.

Rule : 13 Some ***nouns look plural*** but they are ***singular***, most of them are branches of learning, i.e. subjects.

Such as ***mathematics, physics, phonetics, politics, economics, news, innings, mumps, measles, rickets, etc.***

☒ **Wrong** : Mathematics seem to be a difficult subject.

☑ **Right** : Mathematics seems to be a difficult subject.

☒ **Wrong** : The first inning is going on.

☑ **Right** : The first innings is going on.

NOTE

1. Mathematics (plural) = Calculations
2. Politics (plural) = Thoughts
3. Statistics (plural) = Data

- ***My mathematics are not so strong, means my calculations are not so strong.***

Rule : 14 List of ***collective nouns***.

1.	A pile of books.	**21.**	A reel of film.
2.	A set of utensils.	**22.**	A sheaf of grain.
3.	A series of events.	**23.**	A cloud of dust.
4.	A class of pupils.	**24.**	A regiment of soldiers.
5.	A choir of singers.	**25.**	A team of players.
6.	A circle of friends.	**26.**	A gaggle of geese.
7.	A board of directors.	**27.**	A herd of cattle.
8.	A hack of smokes.	**28.**	A hive of bees.
9.	A bevy of girls / women.	**29.**	A troop of scouts.
10.	A cast of actors.	**30.**	A block of flats.
11.	A bunch of crooks.	**31.**	A chest of drawers.
12.	A bouquet of flowers.	**32.**	A pride of lions.
13.	A gang of thieves.	**33.**	A galaxy of stars.
14.	A staff of employees.	**34.**	A fleet of ships.
15.	A cluster of spiders.	**35.**	A bale of cotton.
16.	A brood of hens.	**36.**	A stud of horses.
17.	A crew of sailors.	**37.**	A school of whales.
18.	A crowd of people.	**38.**	A bundle of sticks.
19.	A shoal of fish.	**39.**	A shower of rain.
20.	A swarm of flies.	**40.**	A tribe of natives.

Rule : 15 Some nouns change their meanings when we add *+'s' / + 'es'* to them.

1. Work Works — Factory, literature
2. Alphabet Alphabets — Languages
3. Wood Woods — Small forest
4. Force forces — A group of soldier
5. Water Waters — Sea / ocean
6. Iron Irons — Fetters
7. Copper Coppers — Things which are made of copper
8. Practice Practices — Habits

NOTE

Works, woods, irons etc. look ***plural nouns***, but they are not the plural form of work, wood, iron etc. They must be treated as separate words, because work, wood, iron etc. are uncountable nouns, and uncountable nouns cannot be changed into plural. We need a partitive to do so.

Such as : ***so many pieces of work, two loaves of bread etc.***

Now please do study the following phrases carefully

Sl.	Wrong Phrases	Right Phrases	Sl.	Wrong Phrases	Right Phrases
1.	Angry mob	Mob	11.	Free gift	Gift
2.	Cousin brother / sister	Cousin	12.	Date of birth	Date of the birth
3.	Big blunder	Blunder	13.	Female compartment	Women's compartment
4.	Final result	Result	14.	Lady doctor	Woman doctor
5.	English teacher	Teacher of English	15.	Dressing sense	Dress sense
6.	Family members	Members of family	16.	Passing marks	Pass marks
7.	True fact	Fact	17.	Saving	Savings
8.	Metropolitan cities	Metropolans	18.	Content	Contents
9.	Departmental store	Department store	19.	Past history	History
10.	Saving account	Savings account	20.	Future plan	Plan
			21.	Ticket window	Ticket counter

PRACTICE EXERCISE Nouns

Instructions : Out of the following four sentences, select the most suitable sentence with respect to grammar and usage.

1. (a) I love reading poetries of John Keats.
(b) I love to read poetries of John Keats.
(c) I love reading poems of John Keats.
(d) I love reading poem John Keats.

2. (a) Her hairs are grey.
(b) Her hair are grey.
(c) Her hairs is grey.
(d) Her hair is grey.

3. (a) I love the content of this book.
(b) I love the contents of this book.
(c) I am loving the contents of this book.
(d) I am to love the content of this book.

4. (a) I bought this book from market, and this bar of soap from a departmental store.
(b) I bought this book from market, and this soap from a department store.
(c) I bought this book from the market, and this bar of soap from a department store.
(d) I bought this book from market, and this bar of soap from a department store.

5. (a) I have read two third of this novel.
(b) I have read two thirds of this novel.
(c) I want to read two third of this novel.
(d) I have read two thirds of this novels.

6. (a) One of my friends are coming.
(b) One of my friend are coming.
(c) One of my friends is coming.
(d) One of my friend is coming.

7. (a) I like the sceneries of Kashmir.
(b) I like the scenery of Kashmir.
(c) I likes the scenery of Kashmir.
(d) I likes the sceneries of Kashmir.

8. (a) Raman has sold all the furnitures of his newly constructed office.
(b) Raman has sold every furnitures of his newly constructed office.
(c) Raman has sold all furnitures of his newly constructed office.
(d) Raman has sold all the furniture of his newly constructed office.

9. (a) She wants to buy ten dozens apples.
(b) She wants to buy ten dozen apples.
(c) She wants to buy ten dozens apple.
(d) She wants to buy ten dozen apple.

10. (a) The number of crimes seem to be increasing.
(b) The number of crimes seems to be increasing.
(c) The numbers of crimes seems to be increasing.
(d) The number of crime seems to be increasing.

Answer Key Nouns

1. (c)	2. (d)	3. (b)	4. (d)	5. (b)	6. (c)
7. (b)	8. (d)	9. (b)	10. (b)		

Explanations Nouns

1. (c)

Poetry is an uncountable noun, so it has no plural form, so options (a) and (b) are wrong.

Option (d) makes no sense at all.

Option (c) is the best option choice.

2. (d)

Hair is a singular uncountable noun so it takes singular verb.

The word hairs is wrong, so is option (a).

Option (b) is wrong because it has verb error.

Option (c) also has verb error.

Option (d) is the best option choice.

3. (b)

The proper word here is contents, so options (a) and (d) are wrong.

Option (c) is wrong because of wrong tense.

Option (b) is the best option choice.

4. (d)

The proper words are, a bar of soap and department store, so Option (a) and (b) are wrong.

Option (c) is wrong because article the has been used before market.

Option (d) is the best option choice.

5. (b)

We can say one third, one fourth etc., but we do not say two third, three fourth etc. We should say two thirds, three fourths etc.

According to above rule options (d) incorrect because sentence is talking about one novel.

Option (b) is the best option choice.

6. (c)

The Rule says :

[One of + Plural subject + singular verb]

According to above mentioned rule options (a), (b) and (d) are wrong.

Option (c) is the best option choice.

7. (b)

Scenery is an uncountable noun, so it has no plural form, so options (a) and (d) are wrong.

Option (c) has verb error, so it is also wrong.

Option (b) is the best option choice.

8. (d)

The proper word is furniture, so options (a), (b) and (c) are wrong.

Option (d) is the best option choice.

9. (b)

The correct phrase here should be ten dozen apples, so options (a), (c) and are wrong.

Thus option (b) is the best option choice.

10. (b)

The Rule says :

The number of + Plural noun + Singular verb]

According to rule options (a), (c) and (d) are wrong.

According to rule option (b) is the best option choice.

CHAPTER 4 Pronouns

Pronouns

Pro + Noun

'Pro' means at the place of

A word which is used at the place of a noun or a noun phrase is called pronoun.

Examples –

1. Meera is a good student.
2. She is one of my friends.

   Here *'Meera'* is a ***noun*** *'She'* is a ***pronoun***. 'Meera ' is the antecedent of the pronoun 'She'. Thus we see that - ***I, we, you, he, she, it, they*** are all pronouns.

Types of Pronoun

1. Personal pronouns

Indicate persons such as, ***I, we, you, he, she, it, they***.

2. Relative pronouns

Show relations such as, ***who, which, that, whom, whose***.

3. Reciprocal pronouns

Show relationship with one another such as, ***each other, one another***, etc.

4. Reflexive pronouns

The **singular pronouns** ***myself, yourself, himself, herself, it self*** and the **plural pronouns** ***ourselves, yourselves and themselves*** are called **reflexive pronouns**. They are used when the action done by the subject affects upon the subject.

Examples –

- She is looking at herself in the mirror.
- He cut himself.

5. Indefinite pronouns

Indefinite pronouns are used to denote persons or things in a general way.

The indefinite pronouns are :

- **Singular :**

Somebody, nobody, anybody, everybody, someone, no one, anyone, everyone something, nothing, anything, everything, etc.

- **Plural :**

Both, few, many, several, others, all, etc.

- **Singular and Plural :**

All, any, more, most, none, some, etc.

6. Interrogative pronouns

They are used to ask questions. Interrogative pronouns are

Who, whom, which, what, etc.

Examples –

- Who is that man?
- Which is your pen?

7. Distributive pronouns

Distributive pronouns are used to denote persons or things one at a time. These are

Each, either, neither, etc.

Examples –

- Each of the articles is informative.
- Either of you is not allowed to sit at this desk.

8. Demonstrative pronouns

Demonstrative pronouns are used to point out a thing or things / person / persons these are:

This, that, these, those, etc.

9. Possessive pronouns

Possessive pronouns are used to denote possession or ownership.

Mine, his, hers, theirs, yours, etc.

Points to Remember

- Without pronouns we have to repeat the same nouns over and over again.

  Examples –

  - Tom said that Tom was going to call Tom's father.

    This sentence can be written as :

  - Tom said that he was going to call his father.

- Every pronoun has an antecedent.

- The antecedent is the noun that the pronoun replaces.

- The pronoun must agree with the antecedent in person, case, and gender.
- Personal pronouns have three persons. Such as

1. **First person**	*I, We* (speaker)
2. **Second person**	*You*, (listener)
3. **Third person**	*he, she, it, they,* (The person / persons that we talk about}

- *You* : Singular and Plural.
- *'You'* is used for singular / plural / male / female.
- A 'pronoun' like a noun must be used alone as sub or as an object etc.

	Subjective case	Objective case	Possessive case	Possessive pronoun
First Person	I	me	my	mine
	We	us	our	ours
Second Person	You	you	your	yours
Third Person	He	him	his	his
	She	her	her	hers
	It	it	its	—
	They	them	their	theirs

NOTE

My, our, your, her, their etc. are ***pronomial adjectives*** also known as ***possessive adjective***. They are called possessive adjectives because they cannot be used alone.

Example –

- We cannot say.
- My is coming

We say : My friends are coming.

Rules Regarding Pronouns

Rule : 1 Always use a subjective case after a helping verb, and an objective case after a main verb.

☒ **Wrong** : It is me who can solve these sums.
☑ **Right** : It is I who can solve these sums.

☒ **Wrong** : It was them who were making a noise.
☑ **Right** : It was they who were making a noise.

☒ **Wrong** : Lydia will teach you and he.
☑ **Right** : Lydia will teach you and him.

Rule : 2 Use an objective case after ***'Let'***

☒ **Wrong** : Let you and I compose a song.
☑ **Right** : Let you and me compose a song.

Rule : 3 Use an objective case after a preposition.

☒ **Wrong** : This matter should be between you and I.
☑ **Right** : This matter should be between you and me.

☒ **Wrong** : This problem is so complicated that it cannot be solved by we people.
☑ **Right** : This problem is so complicated that it cannot be solved by us people.

Rule : 4 Use a possessive case before a gerund.

[V_1 + ing – functions as a noun is called gerund]

☒ **Wrong** : Would you mind me sitting here?
☑ **Right** : Would you mind my sitting here?

☒ **Wrong** : Nobody likes them shouting at that mendicant.
☑ **Right** : Nobody likes their shouting at that mendicant.

Rule : 5 When ***'one'*** is used as a pronoun it means people in general including the speaker. ***'one'*** (as a pronoun) is followed by ***one's***.

☒ **Wrong** : One should do his duty.
☑ **Right** : One should do one's duty.

When 'one' is a number do not use one's

☒ **Wrong** : One of them has done one's work.
☑ **Right** : One of them has done his work.

Rule : 6 Reflexive pronouns such as

Myself, himself, yourself, etc. cannot be used as a ***subject*** of a statement.

☒ **Wrong** : Mohan and myself will go home tomorrow to meet them.
☑ **Right** : Mohan and I will go home tomorrow to meet them.

Rule : 7 ***Each other / one another***

'Each other' is used for two people whereas ***one another*** is used for more than two people.

☒ **Wrong** : Ram and Mohan help one another.
☑ **Right** : Ram and Mohan help each other.

☒ **Wrong** : Ram, Mohan and Sohan help each other.
☑ **Right** : Ram, Mohan and Sohan help one another.

Rule : 8 Relative pronouns ***(who, which, that etc.)***

Two sentence can be combined by using a relative pronoun. If the same noun appears in two statements, one of the nouns can be changed into a relative pronoun.

EXAMPLES –

- He likes the car? His father bought the car.

These sentences can be combined into one by using a relative pronoun.

- He likes the car that his father bought.

Usage of relative pronouns

1. ***Who, whom and whose*** are used to replace animate noun (Person)
2. ***Which*** is used to replace non living thing.
3. ***That*** is used to replace animal and non living thing.

EXAMPLES –

- I met a man ***who*** was wearing a pair of blue jeans.
- This is the book ***which*** you gave me.

OR

- This is the book ***that*** you gave me.

When the relative pronouns are in the different cases, one is in the nominative case and the other is in the objective case, they must be mentioned twice.

- The girl ***who*** is my friend and ***whom*** you met in the library yesterday left for London this afternoon.

In case of ***all, same, nothing, the little, the few, the only, superlative degree,*** use ***'that'***

All Same Nothing The only Superlative degree The little The few	+ Who ☒ + Which ☒ + That ☑

☒ **Wrong** : All which glitters is not gold.
☑ **Right** : All that glitters is not gold.

☒ **Wrong** : She is the only one of those writers who has written a number of books on phonetics.
☑ **Right** : She is the only one of those writers that has written a number of books on phonetics.

When there are two antecedents, a man and an animal or two things before the relative pronouns use ***'that'*** at the place of ***who / which***.

☒ **Wrong** : The woman and her cat who was going to Agra injured in an accident.
☑ **Right** : The woman and her cat that was going to Agra injured in an accident.

'Which' is also used for selection. In this case it can be used for persons / things.

☒ **Wrong** : I do not know who of the students broke the chair.
☑ **Right** : I do not know which of the students broke the chair.

Use *whom* after preposition

- To ***whom*** do you want to meet?
- Use ***whose*** for person and of ***which*** for things.

- They visited their grandfather, whose farm is located in Meerut.
- She bought a beautiful hat of which colour was red.

Rule : 9 Usage of ***'all'***

1. Use with countable nouns to mean without exception

Ex. :
- All books must be returned by the 6th June.
- With uncountable nouns to mean the whole of.

I have spent all my money.

As a head word, the meaning of ***'all'*** is close to everybody, everything.

- All agree that he is honest (Everybody agrees ......)
- All is over (everything is over)

Rule : 10 Collective nouns are singular when they designate a group acting as a unit, and they are plural when the members who make up the group are acting independently.

☒ **Wrong** : The band played their fourth concert.
☑ **Right** : The band played its fourth concert.
(Here the band has been used as a single unit)

☒ **Wrong** : The band were tuning its instruments.
☑ **Right** : The band were tuning their instruments.
(Here we are talking about individual members of the band)

Rule : 11 When a pronoun has two antecedents, it should agree with the nearer one.

☒ **Wrong** : I hold in high esteem everything and everybody which reminds me of my attainments.
☑ **Right** : I hold in high esteem everything and everybody who reminds me of my attainments.

Rule : 12 The pronoun ***'as'*** should be used as a relative pronoun in place of who, which, that if same / such comes in the previous clause.

☒ **Wrong** : He is such a fool that I have ever seen.
☑ **Right** : He is such a fool as I have ever seen.

Rule : 13 usage of ***Neither / None / Either***

'Neither' is used for two people in negative sense. ***'None'*** is used for more than two people in negative sense.

☒ **Wrong** : Neither of the ten students was going to attend his class.
☑ **Right** : None of the ten students was going to attend his class.

'Either' is used for two people in positive sense.

- Either of the two boys was playing in the field.

Rule : 14 *Verbs* such as ***enjoy, introduce, resign, apply, acquit, absent, avail, adapt, reconcile, amuse*** etc., are followed by ***reflexive pronouns***.

☒ **Wrong** : She prides on her beauty, so nobody likes her.

☑ **Right** : She prides herself on her beauty, so nobody likes her.

Rule : 15 Reflexive pronouns are not used with the following verbs.

Such as ***keep, conceal, qualify, spread, rest, stay, bathe, move, hide,*** etc.

☒ **Wrong** : You should keep yourself from bad girls.

☑ **Right** : You should keep from, bad girls.

Rule : 16 When first, second and third person singular pronouns (***I, you,*** and ***he***) are used in a sentence together, their sequence should be 231 but in case of crime it should be 123.

☒ **Wrong** : I, you and he went to Agra to see the Taj Mahal.

☑ **Right** : You, he and I went to Agra to see the Taj Mahal.

☑ **Right** : I, you and he killed a lion yesterday.

Rule : 17 In case of plural pronouns the sequence should be 123.

Ex .: We, you and they help the poor on Sundays.

Rule : 18 Usage of *'It'*

'It' is an empty subject. We often use *'it'* in sentences referring to ***time, weather, temperature or distance***. When used in this way, *'it'* is sometimes called ***'empty subject'***, because it carries no real information.

Time : It is 8 o'clock.

Weather : It is raining.

Temperature : It is 0°C. [***Say :*** Zero Degrees Celsius ; ***Don't say :*** Zero Degree Celsius]

Distance : It is 20 miles from London.

We use it for wild animals, birds and baby (whose sex is not known)

☒ **Wrong** : The baby was crying for his mother.

☑ **Right** : The baby was crying for its mother.

We also use *'it'* in cleft sentences.

- It was I who phoned you last night.

It + ***being*** **+** ***day, Sunday***
+ ***holiday***
+ ***weather***
+ ***month*** etc. apply

It before being.

☒ **Wrong** : Being Sunday I could not meet one of my relatives.

☑ **Right** : It being Sunday, I could not meet one of my relatives.

If there is ***an adjective*** after ***being***, do not use *'it'* before ***being***.

☒ **Wrong** : Being poor, I admitted him to hospital.

☑ **Right** : He being poor, I admitted him to hospital.

Rule : 19 Use of Pronoun after ***than / as***

If there is a comparison between two subjects we use subjective case after ***than / as***.

☒ **Wrong** : He is as slow as me.
☑ **Right** : He is as slow as I am.

☒ **Wrong** : You are faster than him.
☑ **Right** : You are faster than he is.

Rule : 20 If there is a difference in the tense of the main clause and the ***(than / as)*** clause. The helping verb of the tense must be used.

1. I learn more than he will.
2. He knows more than I did at his age.

Rule : 21 When each comes after a plural noun / pronoun, use plural verb / plural pronoun.

☒ **Wrong** : The students each completed his home work.
☑ **Right** : The students each completed their homework.

PRACTICE EXERCISE **Pronoun**

Instructions : Out of the following four sentences, select the most suitable sentence with respect to grammar and usage.

1. (a) I do not like them coming late.
(b) I do not like they coming late.
(c) I do not like their coming late.
(d) I do not like they came late.

2. (a) This matter should be between you and I.
(b) This matter should be between I and you.
(c) This matter should be between me and you.
(d) This matter should be between you and me.

3. (a) It is me who can solve all the sums.
(b) It is I whom can solve all the sums.
(c) It is me that can solve all the sums.
(d) It is I who can solve all the sums.

4. (a) Let you and he go home with them.
(b) Let he and you go home with them.
(c) Let you and him go home with them.
(d) Let him and you go home with them.

5. (a) Being Sunday, I did not go to office.
(b) It been Sunday, I did not go to office.
(c) It being Sunday, I did not go to office.
(d) It has being Sunday, I did not go to office.

6. (a) I saw a man which was siting on the table.
(b) I saw a man whom was sitting on the table.
(c) I saw a man who were sitting on the table.
(d) I saw a man who was sitting on the table.

7. (a) The baby was crying for his mother.
(b) The baby was crying for her mother.
(c) The baby was crying for their mother.
(d) The baby was crying for its mother.

8. (a) I, you and he will go to school.
(b) He, I and you will go to school.
(c) You, he and I will go to school.
(d) You, I and he will go to school.

9. (a) One should do their duty.
(b) One should do one's duty.
(c) One should do his duty.
(d) One should do her duty.

10. (a) The master did not know who of the students broke the chair.
    (b) The master did not know which of the student broke the chair.
    (c) The master did not know which of the students broke the chair.
    (d) The master did not know which of the students broken the chair.

Answer Key ➤ Pronoun

1. (c)	2. (d)	3. (d)	4. (c)	5. (c)	6. (d)
7. (d)	8. (c)	9. (b)	10. (c)		

Explanations ➤ Pronoun

1. **(c)**

The Rule says :

Possessive case + Gerund (V_1 + ing)

According to rule there should be their before coming, so options (a), (b) and (d) are wrong.

Option (c) is the best option choice.

2. **(d)**

The Rule says :

Preposition + Objective case of a pronoun

According to rule options (a) and (b) are wrong.

Option (c) is wrong because ***me*** and ***you*** cannot be used.

Option (d) follows the above mentioned rule so option (d) is the best option choice.

3. **(d)**

After helping verb we apply subjective case in formal English, so option (a) is wrong.

Here the objective pronoun ***'whom'*** makes the option (b) wrong.

Option (c) is wrong because ***that*** is wrong.

Thus option (d) is the best option choice.

4. **(c)**

The Rule says :

Let + Objective case + V_1 ....

According to the above mentioned rule options (a) and (b) are wrong.

Him and ***You*** cannot be used, so option (d) is wrong.

5. **(c)**

The proper phrase is, It being Sunday, so options (a), (b) and (d) are wrong.

Options (c) is the best option choice.

6. **(d)**

For person we use ***who***, so options (a) and (b) are wrong.

Option (c) is wrong because it has verb error.

Option (d) is the best option choice.

7. **(d)**

Baby (whose sex is not known) is considered neuter in English, so options (a), (b) and (c) are wrong.

Option (d) is the best option choice.

8. **(c)**

If all the three persons are given in a sentence, their sequence should be **2 3 1** but incase of crime it should be **1 2 3**.

According to the above mentioned rule options (a), (b) and (d) are wrong.

Option (c) is the best option choice.

9. **(b)**

One should be followed by ***one's***, so options (a), (c) and (d) are wrong.

Option (c) is the best option choice.

10. **(c)**

The correct phrase should be ***which of the students*** so options (a) and (b) are wrong.

Option (d) is wrong because it has verb error.

Thus, option (c) is the best option choice.

CHAPTER 5

Adjectives

Adjectives

An adjective is a word that modifies, or describes a noun or a pronoun.

EXAMPLES —

1. The clouds are heavy and dark.
2. Candida is a beautiful woman.

Kind of Adjectives

1. Adjectives of quality

Tell quality of persons, places, or things

Bad, good, clever, great, handsome, intelligent, etc.

EXAMPLES —

- Sandra is an intelligent student.

2. Adjectives of quantity

Tell quantity of things

Many, much, a great deal of , a lot of, few, little, etc.

EXAMPLES —

1. I have ***many*** books to read.
2. The teacher could not teach because there were ***few*** students in the classroom.

3. Adjectives of number

Tell how many persons, or things are meant.

One, two, three, ... first, second, third, etc.

EXAMPLES —

1. They bought ten novels yesterday.
2. The first three chapters of this book are difficult.

4. Demonstrative adjectives

These adjectives point out persons / things.

This, that, these, those, such, the same, etc.

EXAMPLES —

1. I like this book.
2. Those boys were going to school.

5. Distributive adjectives

These adjectives show that persons or things are taken separately.

Each, every, either, neither, etc.

EXAMPLES —

1. Each girl will get a bar of chocolate.
2. One of my friends goes to temple every day.

6. Proper adjectives

Adjectives that come from proper nouns are called proper adjectives

EXAMPLES —

1. Chirag is an Indian player.
2. A group of Australian students wants to join our team.

7. Interrogative adjectives

They are used for asking questions.

which, whose, what, etc.

EXAMPLES —

1. Which book do you want to read?
2. What time would you like to meet me?

8. Possessive adjectives

They are used to show ownership.

my, his, her, your, their, etc.

EXAMPLES —

1. Your girlfriend seems to be stunning.
2. My house is better than yours.

9. Exclamatory adjectives

What, how, etc.

EXAMPLES —

1. What a beautiful the building is!

Adjectives can be used in two ways

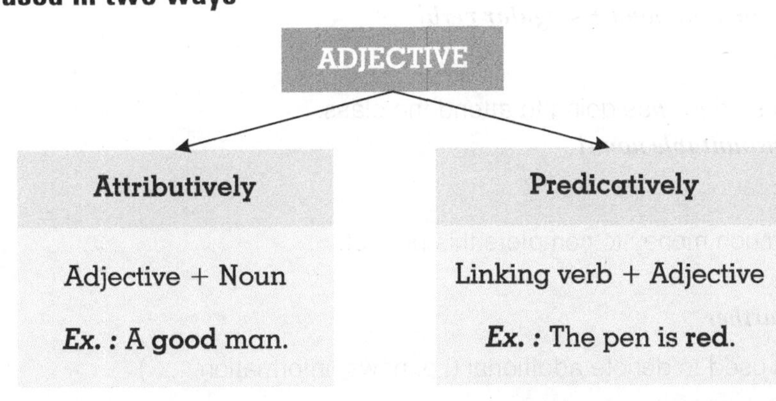

- When an ***adjective*** comes before a noun it is said to be attributive use of an adjective.
- When an ***adjective*** is used after a ***linking verb*** such as ***is/am/are/was/were/feel/smell/taste, sound,*** etc. it is said to be predicative use of an adjective.

☒ **Wrong** : These mangoes taste bitterly ***(Adverb)***.
☑ **Right** : These mangoes taste bitter ***(Adjective)***.
[Taste (a link verb) needs an adjective not an adverb]

Rules regarding Adjectives

Rule : 1 *Some/Any*

'Some' is used with positive statements.
'Any' is used with negative and interrogative statements.

EXAMPLES —

1. I have some novels for Mohan.
2. Do you have any pen?
3. I do not have any pen to write with.

> *'Some'* can be used with ***interrogative statement*** if answer is expected ***positively***.

EXAMPLES —

1. Would you like to eat something?
2. Can you lend me some money?

☒ **Wrong** : Somebody hardly gave him something to eat.
☑ **Right** : Somebody hardly gave him anything to eat.

Rule : 2 *Many/Much/Many a/A good many/A great many.*

[Many, A good many, A great Many] + Plural countable noun]

EXAMPLE —

- Many people were killed in that accident.

Rule : 3 *[Many a + singular noun + singular verb]*

EXAMPLE —

- Many a student was going to attend the class.

[Much + uncountable noun]

EXAMPLE —

- I want much money to complete this project.

Rule : 4 *Further/Farther*

'Further' is used to denote additional (i.e. news, information ......)
'Farther' is used to denote distance.

☒ **Wrong** : He will provide you farther information regarding this case.
☑ **Right** : He will provide you further information regarding this case.

Rule : 5 ***Nearest / Next***

'Nearest' refers to something that is at a shortest distance away.

'Next' denotes the one of a sequence of things that come after the other.

EXAMPLE —

1. The nearest bus stop is 2 kms away.
2. The book shop is next to the theatre.

Rule : 6 ***Former, Latter, First, Last and Later***

Use ***'former'*** and ***latter*** when you discuss two items of a series.

EXAMPLE —

1. Ram and Mohan are my friends but I like the former [Ram]

   We use ***'first'*** and ***'last'*** when we talk about the series of more than two people / things.

EXAMPLE —

1. ***'A'*** is the first vowel of English alphabet.
2. ***'U'*** is the last vowel of English alphabet.

   Use later when you talk about time.

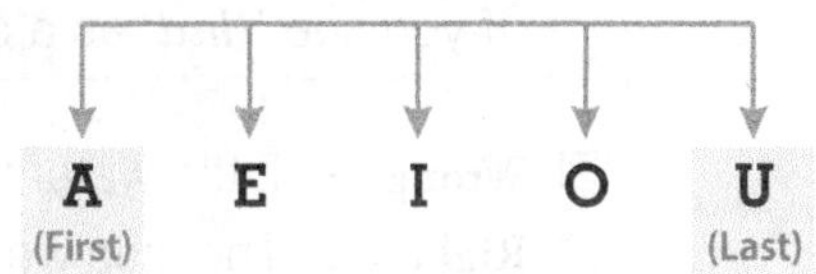

EXAMPLE —

1. Would you mind my explaining this chapter later?
2. I will talk to you later.

Rule : 7 **A.** ***Few, A few, The few***

1. ***'Few'*** means hardly any [Gives negative sense]
2. ***'A few'*** means not many [Gives positive sense]
3. ***'The few'*** means at least some [Gives positive sense]

   Few, ***a few*** and ***the few*** are used before plural countable noun.

   [Few, a few and the few + plural countable noun]

☒ **Wrong** : I have few book.
☑ **Right** : I have few books.

☒ **Wrong** : I have few books, so you can take two of them.
☑ **Right** : I have a few books, so you can take two of them.

☒ **Wrong** : She gave mendicant a few coins that she had in her purse.
☑ **Right** : She gave mendicant the few coins that she had in her purse.

> **NOTE**
>
> If you see ***'that'*** as a conjunction in the statement use ***'the'*** before few.

☒ **Wrong** : Few books that you gave me were interesting.
☑ **Right** : The few books that you gave me were interesting.

Rule : 7 B. ***Little, A little, The little***

'Little' means almost nothing [Gives negative sense]

'A little' means not much [Gives positive sense]

'The little' mans at least some [Gives positive sense]

Little, a little and the little are used before uncountable noun.

[Little, a little and the little + uncountable noun]

☒ **Wrong** : Rochester was very thirsty, so Cathy gave him little water to drink.
☑ **Right** : Rochester was very thirsty, so Cathy gave him a little water to drink.

☒ **Wrong** : She could not prepare tea because there was a little milk in the pot.
☑ **Right** : She could not prepare tea because there was a little milk in the pot.

☒ **Wrong** : Little knowledge that he got from newspapers and magazines was not sufficient for him even to get pass marks in his exams.
☑ **Right** : The little knowledge that he got from newspaper and magazine was not sufficient for him even to get pass marks in his exams.

NOTE

If you see *'that'* as a conjunction in the statement use *'the'* before little.

☒ **Wrong** : Little water that was in the bucket, evaporated.
☑ **Right** : The little water that was in the bucket, evaporated.

Rule : 7 C. ***No less than / No Fewer than***

'No less than' is used before uncountable noun.

'No fewer than' is used before plural countable noun.

☒ **Wrong** : No less than forty people were invited to the party.
☑ **Right** : No fewer than forty people were invited to the party.

☒ **Wrong** : There were less people in that department store yesterday.
☑ **Right** : There were fewer people in that department store yesterday.

Rules Regarding Comparisons

Descriptive adjectives have three degrees of comparison.

1. Positive degree
2. Comparative degree
3. Superlative degree

EXAMPLES —

Positive Degree	Comparative Degree	Superlative Degree
Good	Better	Best
Small	Smaller	Smallest
Hot	Hotter	Hottest
Beautiful	More beautiful	Most beautiful
Intelligent	More intelligent	Most intelligent

Rule : 1 Do not use double comparative or superlative degree of an adjective simultaneously (together).

☒ **Wrong** : Sandra is more wiser than Alison.
☑ **Right** : Sandra is wiser than Alison.

☒ **Wrong** : Joseph is the most cleverest student in my class.
☑ **Right** : Joseph is the cleverest student in my class.

Rule : 2 When two qualities of a person are compared use more + positive degree of an adjective.

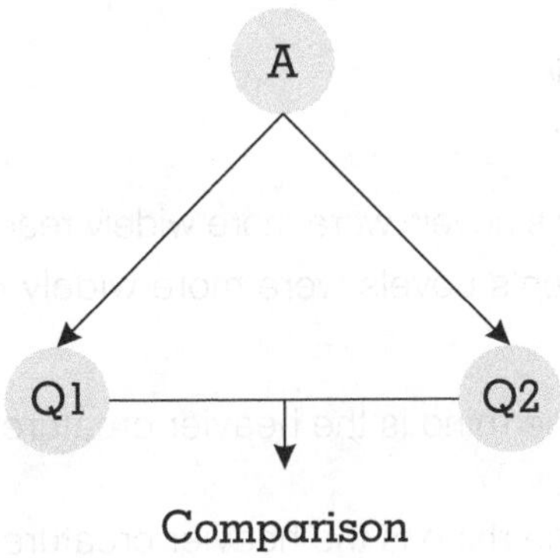

More + Positive degree of an adjective

i.e. more + wise

☒ **Wrong** : Meera is wiser than honest.
☑ **Right** : Meera is more wise than honest.

Rule : 3 The adjectives that end in *'or'* take the preposition *'to'* not *'than'*.

Junior, superior, senior, inferior etc.

☒ **Wrong** : Sheela is senior than me.
☑ **Right** : Sheela is senior to me.

☒ **Wrong** : His work is inferior than mine.
☑ **Right** : His work is inferior to mine.

Rule : 4 Prefer

Prefer is followed to, not than

☒ **Wrong** : I prefer reading than writing.
☑ **Right** : I prefer reading to writing.

Rule : 5 Some adjectives are nongradable, so do not use more / most before them.

entire, ideal, absolute, supreme, universal, unique, perfect, square, excellent, complete, etc.

☒ **Wrong** : Lydia is the most perfect actor for this movie.
☑ **Right** : Lydia is a perfect actor for this movie.

Rule : 6 Comparative degree is used to compare two people / things of same class.

Person – Person
Thing – Thing
Sub – Sub

☒ **Wrong** : My book is better than you.
☑ **Right** : My book is better than yours.

☒ **Wrong** : The climate of Agra is hotter than Shimla.
☑ **Right** : The climate of Agra is hotter than that of Shimla.

☒ **Wrong** : She is wiser than me.
☑ **Right** : She is wiser than I.
OR
She is wiser than I am.

☒ **Wrong** : This question is the easiest of the two.
☑ **Right** : This question is the easier of the two.

☒ **Wrong** : In Victorian England, Charles Dicken's novels were more widely read than Victor Hugo.
☑ **Right** : In Victorian England, Charles Dicken's novels were more widely read than those of Victor Hugo.

☒ **Wrong** : Between the rhino and the hippo, the rhino is the heavier creature while the hippo is the most ferocious.
☑ **Right** : Between the rhino and the hippo, the rhino is the heavier creature while the hippo is the more ferocious.

Rule : 7 We use superlative degree when we compare three or more things.

☒ **Wrong** : The executive interviewed five candidates for the post of editor and ultimately decided that Manju was the more qualified.
☑ **Right** : The executive interviewed five candidates for the post of editor and ultimately decided that Manju was the most qualified.

☒ **Wrong** : This question is the easier of the three.
☑ **Right** : This question is the easiest of the three.

Rule : 8 If two qualities of a person or a thing are defined, they should be in the same degrees of comparison.

☒ **Wrong** : Orlando is the most intelligent and honest student of my class.
☑ **Right** : Orlando is the most intelligent and most honest student of my class.

Rule : 9 When two things are compared, are equal we use *'as .... as'* in a positive statement, and *so ...... as* in a negative statement.
Use positive degree of an adjective between *'as ..... as'* / *so ..... as*.

☒ **Wrong** : She is as wiser as Meena.
☑ **Right** : She is as wise as Meena.

☒ **Wrong** : Mohan is as miser as Deepak.
☑ **Right** : Mohan is as miserly as Deepak.
***Miser* (noun) *Miserly* (Adjective)**

Rule : 10 Always use positive degree of an adjective after *'Comparatively'* and *'relatively'*.

☒ **Wrong** : She is comparatively better today.
☑ **Right** : She is comparatively well today.

PRACTICE EXERCISE Adjectives

Instructions : Out of the following four sentences, select the most suitable sentence with respect to grammar and usage.

1. (a) I am waiting for the revise edition of this book.
(b) I am waiting for the revised edition of this book.
(c) I am waiting for the revising edition of this book.
(d) I am waiting for the revises edition of this book.

2. (a) My house is three times bigger than Sohan's.
(b) My house is three times bigger than Sohan's house.
(c) My house is three times as big as Sohan's
(d) My house is three times as big as Sohan.

3. (a) Raman is as miser as Ramesh.
(b) Raman is mise as Ramesh.
(c) Raman is as miserly as Ramesh.
(d) Raman is miser as Ramesh.

4. (a) These mangoes taste sweeter.
(b) These mangoes taste sweetly.
(c) These mangoes taste sweet.
(d) These mangoes tastes sweet.

5. (a) Radha is the most unique singer in my city.
(b) Radha is a most unique singer in my city.
(c) Radha is the unique singer in my city.
(d) Radha is a unique singer in my city.

6. (a) He is wise then me.
(b) He is wise than me.
(c) He is wiser than I am.
(d) He is more wiser than me.

7. (a) Ram is the most intelligent and honest student in my class.
(b) Ram is the intelligent and most honest student in my class.
(c) Ram is the most intelligent and most honest student in my class.
(d) Ram is the most intelligent and the most honest student in my class.

8. (a) Many a student has gone to school.
(b) Many a students have gone to school.
(c) Many student has gone to school.
(d) Many a students has gone to school.

9. (a) Madan is wiser than handsome.
(b) Madan is more wise than handsome.
(c) Madan is most wise than handsome.
(d) Madan is more wiser than handsome.

10. (a) I prefer reading than writing.
(b) I prefer reading over writing.
(c) I prefer reading with writing.
(d) I prefer reading to writing.

Answer Key ➤ Adjectives

1. (b) **2.** (c) **3.** (c) **4.** (c) **5.** (d) **6.** (c)
7. (c) **8.** (a) **9.** (b) **10.** (d)

Explanations ➤ Adjectives

1. (b)

Option (a) is wrong because revise edition is a wrong phrase.

Option (c) is wrong because revising edition is a wrong phrase.

Option (d) is wrong because revises edition is a wrong phrase.

Option (b) is correct because ***revised edition*** is a correct phrase.

2. (c)

Three times bigger is wrong, so options (a) and (b) are wrong.

In option (d) my house has been compared to Sohan, so option (d) is wrong.

Option (c) is the best option choice.

3. (c)

We use positive degree of an adjective between as _____ as.

Miser is a noun and miserly is an adjective

According to above mentioned explanation :

Option (a), (b) and (d) are wrong.

Option (c) is the best option choice.

4. (c)

We apply an adjective after taste, so option (a) and (b) are wrong.

Option (d) is wrong because it has subject verb agreement.

Thus, option (c) is the best option choice.

5. (d)

Unique is a non-gradable adjective, it has no comparative and superlative degree.

According to above mentioned explanation option (a) and (b) are wrong.

Option (c) is wrong because article 'the' has been used wrongly before unique.

Thus, option (d) is the best option choice.

6. (c)

To make comparison, we use comparative degree, so option (a) is wrong.

Option (b) is wrong because of wrong comparison.

Double comparative degree is not used, so option (d) is wrong.

Thus, option (c) is the best option choice.

7. (c)

When we define two qualities of a person, same degrees of comparison should be used.

According to above mentioned explanation options (a), (b) and (d) are wrong.

Thus, option (c) is the best option choice.

8. (a)

The Rule says :

Many a + Singular subject + Singular verb

According to above mentioned rule options (b), (c) and (d) are wrong.

Thus, option (a) is the best option choice.

9. (b)

When we compare two qualities of a person, we use ***[More + Positive degree]***.

According to above mentioned explanation options (a), (c) and (d) are wrong.

Thus, option (b) is the best option choice.

10. (d)

Prefer is followed by ***to***.

According to above mentioned rule options (a), (b) and (c) are wrong.

Thus, option (d) is the best option choice.

CHAPTER 6

Adverbs

Adverbs

Adverbs are used to describe, verbs (except linking verbs), adjectives, or other adverbs.

EXAMPLES —

1. He runs fast.

   Here, ***'fast'*** describes the verb ***'run'***.
2. Belinda seems to be a very beautiful woman.

   Here, ***'very'*** describes the adjective ***'beautiful'***.
3. He runs very fast.

   Here ***'very'*** describes another adverb ***'fast'***.

Here we can say ***'An adverb'*** is a word which answers the questions ***How?***, ***When?***, or ***Where?***

Types of Adverbs

1. Adverbs of Manner:

Kindly, well, fast, carefully, beautifully, badly, slowly etc.

An adverb of manner tells us how or in what manner an action is performed.

EXAMPLES —

1. Jane walks gracefully.
2. Raman sang beautifully.
3. I speak Russian slowly.

2. Adverbs of place:

Up, abroad, here, inside, outside, everywhere, somewhere, nowhere, there, home, etc.

An adverb of place tells us where something happens.

EXAMPLES —

1. One of my friends is going abroad.
2. Cathy went somewhere.

3. Adverbs of time:

Yesterday, now, today, yet, tomorrow, recently, at once, since, then etc.

An adverb of time tells us when something happens.

EXAMPLES —

1. He has not eaten food yet.
2. We are moving today.
3. Would you mind my meeting you tomorrow?

4. Adverbs of frequency:

Never, rarely, seldom, occasionally, sometimes, often, usually, frequently, always, rarely.

An adverb of frequency describes ***'How often'*** an action happens.

An adverb of frequency can begin a sentence, precede the main verb, follow be, or end a sentence.

EXAMPLES —

1. I always go to the same cafe for lunch.
2. He has frequently stayed late.
3. He often surfs the web instead of working.
4. Rochester never works on Sundays.

5. Adverbs of opinion:

Personally, obviously, frankly, certainly, luckily, actually, probably, definitely, surely etc.

An adverb of opinion expresses the speaker's opinion.

EXAMPLES —

1. Frankly, I do not like their coming late.
2. You obviously like her winsome smile.

6. Adverbs of degree:

Quite, too, so, very, fairly, hardly, barely, enough, nearly, really, even, etc.

An adverb of degree generally modifies an ***'adj'*** or ***'adverb'*** and indicates intensity (degree)

EXAMPLES —

1. The shoes are too wide.
2. The horses galloped very fast.
3. Neetu is kind enough to help you.

7. Interrogative adverbs:

When? Where? Why? How? An interrogative adverb is used in forming a question.

EXAMPLES —

1. Where do you go with them? **[Come crying = born]**
2. When did she come crying?
3. How do you spell your name?

8. Adverbs of affirmation and negation:

Such as surely, obviously, not, truly, apparently, etc., indicate the certainty of actions, whether they will take place or not.

Rules Regarding Adverbs

Rule : 1 Do not use ***an adverb*** after ***linking verbs***.

☒ **Wrong :** These mangoes taste sweetly.
☑ **Right :** These mangoes taste sweet.

☒ **Wrong :** You look sadly today.
☑ **Right :** you look sad today.

Linking Verbs

is, am, are, was, were, feel, smell, taste, sound, appear, seem, turn, grow, look, etc.

Rule : 2 Do not use *'an adverb'* before a *noun*.

☒ **Wrong :** Please do read this novel thoroughly if you want to have thoroughly knowledge of 'The Victorian period'.

☑ **Right :** Please do read this novel thoroughly if you want to have through knowledge of 'The Victorian Period'.

Explanation - *'Thoroughly'* is an adverb and *'knowledge'* is a noun, so thoroughly cannot be used before knowledge.

The use of *'adjective'* 'through' before' knowledge' is correct.

REMEMBER

1. *Noun + ly = Adjective* 2. *Adjective + ly = Adverb*

Rule : 3 With a view to making our meaning clear, an *'adverb'* should be used close to the word it modifies.

☒ **Wrong :** They only have two months to complete this project.

☑ **Right :** They have only two months to complete this project.

☒ **Wrong :** Candida just wants to buy one book.

☑ **Right :** Candida wants to buy just one book.

☒ **Wrong :** This car nearly costs forty thousand rupees.

☑ **Right :** This car costs nearly forty thousand rupees.

Rule : 4 Adverbs of frequency such as ***always***, ***often***, ***never***, ***ever***, ***usually***, ***seldom*** etc. should be used aftere

(i) the subject

(ii) If a sentence has an ***auxiliary verb*** (helping verb), use it after the ***helping verb***.

Examples:

I often go home to meet my brother.

He will never write a letter to his friends.

Rule : 5 Adjectives that end in by such as ***friendly***, ***lovely***, ***miserly***, ***cowardly***, ***scholarly***, ***niggardly***, ***motherly***, ***fatherly*** etc. cannot be used as adverbs to modify verbs. The following examples will help you make out the correct usage.

☒ **Wrong :** One of my friends behaves friendly.

↓ behaves → ***Verb***; friendly → ***Adjective***

☑ **Right :** One of my friends behaves in a friendly manner

[in + a noun/adjective + ly + manner / way / fashion]

☒ **Wrong :** A soldier is never taught to fight cowardly.

☑ **Right :** A soldier is never taught to fight in a cowardly manner.

Rule : 6 **Too / Very**

'Too' means beyond the proper limit, it indicates unpleasant sense.
'Very' means in a great degree, it indicates pleasant sense.

☒ **Wrong :** I am *too* glad to meet you.

☑ **Right :** I am *very* glad to meet you.

But we can say

The lion was *too* old to hunt the animals.
This can be written as,
The lion was so old that it could not hunt the animals.

Rule : 7 **Too much / Much too**

'Too much' is an adjective, so it is used before uncountable noun.
'Much too' is an adverb, so it is uesd before an adjective.

☒ **Wrong :** *Much too* money is required to complete this project.
☑ **Right :** *Too much* money is required to complete this project.
☒ **Wrong :** His story seems to be *too much* painful.
☑ **Right :** His story seems to be *much too* painful.

Remember

Too much + pain
Much too + painful

Rule : 8 **Fairly / Rather**

'Fairly' (adverb) means to some extent but not very, *Fairly* is used before an adjective/adverb that indicates pleasant sense.

NOTE

1. Fairly can take indefinite article.
   - a fairly pretty women.
   - a fairly lovely child.
2. Fairly is not followed by *too*. Do not say
   - fairly *too* kind.
   - fairly *too* beautiful.

'Rather' means to some degree.
'Rather' is used when we express slight criticism, disappointment or surprise. We can say rather is used before the *adjective / adverb*, that indicates unpleasant sense.

- The instructions were rather complicated.
- She fell and hurt her leg rather badly.

☒ **Wrong :** Sheela is *rather* beautiful.
☑ **Right :** Sheela is *fairly* beautiful.

Rule : 9 **Enough**

'Enough' is both an adjective and an adverb. *Enough* means sufficient.
As an *adjective 'enough'*, is used before *plural* or uncountable *noun* to mean as many as or as much as somebody needs or wants.

- Is there enough room for me?
- Have you got enough pens?

When *'Enough'* is an adverb it goes after adjectives and adverbs.

☒ **Wrong :** She is enough kind to help you.
☑ **Right :** She is kind enough to help you.

☒ **Wrong :** You will not pass the exams if you do not work enough hard.
☑ **Right :** You will not pass the exams if you do not work hard enough.

Rule : 10 Hard / Hardly

'Hard' (adverb) means done with a great deal or force or strength.

Ex. : They work hard at school.

'Hardly' (adverb) means almost not at all.

☒ **Wrong** : The winter months would deal ***hard*** with many of these poor folk.

☑ **Right** : The winter months would deal ***hardly*** with many of these poor folk.

☒ **Wrong** : Meera works ***hardly*** because she wants to get good marks.

☑ **Right** : Meera works ***hard*** because she wants to get good marks.

> **NOTE**
> Use ***'hard'*** when you mean positive sense; use ***hardly*** when you mean negative sense.

Rule : 11 Please / Kindly should not be used simultaneously.

☒ **Wrong** : ***Please kindly*** ask him to take this medicine when he has eaten food.

☑ **Right** : ***Please*** ask him to take this medicine when he has eaten food.

OR ***Kindly*** ask him to take this medicine when he has eaten food.

☑ **Right** : Sandra is fond of ***kindly*** man. (Means Sandra is fond of friendly man.)

> ***Kindly*** is also an ***adjective*** and it is used, to mean ***friendly***.

Rule : 12 If / supposing should not be used simultaneously.

☒ **Wrong** : **If** ***supposing*** a ghost appears in your room, what will you do?

☑ **Right** : **If** a ghost ***appears*** in your room, what will you do?

Rule : 13 Very / Much

Very (adverb) means to a great degree.

Very is used before

- Positive degree of an adjective / adverb

  Ex. : The situation is very serious.
- Present participle ***(i.e. V_1 + ing used as an adjective)***

  Ex. : This novel is very interesting.
- Superlative degree to add force.

  Ex. : This is the very best novel I have ever read.

> **NOTE**
> ***'Very'*** can be used as an adjective.

Much (adverb) means to a great degree or extent.

Much is used before

(i) comparative degree of adjective / adverb

(ii) past participle.

EXAMPLES —

- I am feeling much better.
- I was much exhausted yesterday, so I did not attend the meeting.

☒ **Wrong** : Her story seems to be ***much*** interesting.

☑ **Right** : Her story seems to be ***very*** interesting.

☒ **Wrong** : I feel ***very*** better to see her wearing a mild smile on her rosy lips.

☑ **Right** : I feel ***much*** better to see her wearing a mild smile on her rosy lips.

Rule : 14 Yet / Still

'Yet' (adverb) means so far, until now.

We often use ***'yet'*** with the present perfect tense, in questions and negative statements.

EXAMPLES —

- Why have you not completed your homework ***yet***?
- We have not eaten food ***yet***.

'Still' (adverb) means up to this or that time.

The adverb of duration still goes before the main verb. ***Still*** can go before or after *'be'*.

EXAMPLES —

- He ***still*** lives in Delhi.
- I am ***still*** learning English.

Rule : 15 Else is followed by 'but'

☒ **Wrong** : It is nothing ***else*** then sheer foolishness.

☑ **Right** : It is nothing ***else*** but sheer foolishness.

Rule : 16 Adverbs ***seldom, never, hardly, scarcely, rarely, barely*** etc. have negative meaning, so we do not use any other negative word with them.

☒ **Wrong** : I ***hardly*** know nothing about that stupid woman.

☑ **Right** : I ***hardly*** know anything about that stupid woman.

Rule : 17 If the adverbs, ***never, seldom, no sooner, hardly, scarcely, not only***, etc., come at the beginning of the sentence, use inversion. (i.e. put helping verb before the subject).

☒ **Wrong** : ***Hardly*** he had reached the station, when it began to rain.

☑ **Right** : ***Hardly*** had he reached the station when it began to rain.

☒ **Wrong** : ***Seldom*** he goes to school on time.

☑ **Right** : ***Seldom*** does he go to school on time.

☒ **Wrong** : ***No sooner*** he had entered the office than the boss began to shout.

☑ **Right** : ***No sooner*** had he entered the office than the boss began to shout.

Rule : 18 Some phrases ***'seldom or never'***, ***'seldom if ever'*** little or nothing, and little if anything are wrongly used as ***seldom or ever***, ***seldom if never***, ***little or anything*** and ***little if nothing respectively***.

☒ **Wrong** : They ***seldom*** or ***ever*** come to meet me nowadays.

☑ **Right** : They ***seldom*** or ***never*** come to meet me nowadays.

☑ **Right** : They ***seldom*** if ***ever*** come to meet me nowadays.

Rule : 19 Comparisons with As .... As ....

When two things are compared, are equal, we use ***as* ..... *as* ....**

EXAMPLE —

- John ran ***as*** quickly ***as*** Cathy.

☒ **Wrong** : I did my work ***as*** better ***as*** I could do.

☑ **Right** : I did my work ***as*** well ***as*** I could do.

Rule : 20 Good / Well

'Good' is a true adjective which is used to define the characteristics of a person.

'Well' is both an adjective and an adverb. ***Well*** when used as an adverb means in good health, and ***when*** used as an adjective, means in a good manner.

1. She is ***good***. [She had good behaviour]
   [Adjective]
2. She is ***well***. [She is in good health]
   [Adjective]
3. She speaks ***well***. [She speaks in a good manner]
   [Adverb]

Rule : 21 Some / Somewhat

'Some' is an adjective of indefinite number, and ***'Somewhat'*** is an adverb.

☒ **Wrong** : She is ***some*** better today.

☑ **Right** : She is ***somewhat*** better today.

Clarification : An adverb is required to modify the adjective ***'better'***.

Rule : 22 Only to

'Only to' means much.

EXAMPLE —

- I am only too glad to be here.

Rule : 23 Home

When ***'home'*** is used as an adverb of place.

Neither a preposition nor a possessive adjective should be used before ***home***.

EXAMPLES —

☒ **Wrong** : I am going to home.

☑ **Right** : I am going home.

☒ **Wrong** : I am going my home.

☑ **Right** : I am going home.

PRACTICE EXERCISE Adverb

Instructions : Out of the following four sentences, select the most suitable sentence with respect to grammar and usage.

1. (a) She works as good as she can do.
(b) She works as better as she can do.
(c) She works as well as she can do.
(d) She works as best as she can do.

2. (a) Shyam behaves friendly.
(b) Shyam behaves friendly manner.
(c) Shyam behaves in a friendly manner.
(d) Shyam behaves in friendly fashion.

3. (a) I am too glad to meet you.
(b) I am very glad to meet you.
(c) I am much glad to meet you.
(d) I am too much glad to meet you.

4. (a) I was very astonished to see her smiling.
(b) I was much astonished to see her smiling.
(c) I was too much astonished to see her smiling.
(d) I was very astonish to see her smiling.

5. (a) A soldier is never taught to fight cowardly.
(b) A soldier is never taught to fight in a cowardly manner.
(c) A soldier is never taught to fight in a coward way.
(d) A soldier is never taught to fight in cowardly manner.

6. (a) He works hardly because he wants to get good marks.
(b) He works very hardly because wants to get good marks.
(c) He works scarcely because he wants to get good marks.
(d) He works hard because he wants to get good marks.

7. (a) Her story seems to be too much painful.
(b) Her story seems to be much too painful.
(c) Her story seems to be nearly painful.
(d) Her story seems to be much painful.

8. (a) Never Sohan comes on time to attend the class.
(b) Never Sohan come on time to attend the class.
(c) Never does Sohan come on time to attend the class.
(d) Never does Sohan comes on time to attend the class.

9. (a) One of my friends seems to be sadly today.
(b) One of my friends seems to be in a sadly manner today.
(c) One of my friends seems to be sad today.
(d) One of my friends seems to be in a sadly fashion.

10. (a) This man is enough kind to help you.
(b) This man is kind enough to help you.
(c) This man is enough kindly to help you.
(d) This man is kind enough too help you.

Answer Key ➤ Adverb

1. (c)	2. (c)	3. (b)	4. (b)	5. (b)	6. (d)
7. (b)	8. (c)	9. (c)	10. (b)		

Explanations ➤ Adverb

1. (c)

If there is a verb before ***as .... as***, there should be an adverb between ***as .... as***. Here we see that ***good*** is an adjective and ***well*** is an adverb, so there must be well between ***as .... as***.

According to the above explanation options (a), (b) and are wrong.

Thus, option (c) is the best option choice.

2. (c)

Friendly is an adjective so it cannot be used after the verb behave, thus option (a) is wrong.

Option (b) is wrong because friendly manner is a wrong phrase, so is option (d).

Thus, option (c) is the best option choice.

3. (b)

The adverb too gives unpleasant sense, so options (a), (c) and (d) are wrong.

Thus, option (b) is the best option as we know that ***very*** gives pleasant sense.

4. (b)

We use very before positive degree of an adjective or adverb, whereas ***much*** is used before comparative degree of an adjective or adverb.

Examples : ***Very good***
Much better
Much astonished

Now we can easily conclude that options (a), (c) and (d) are wrong.

Thus option (b) is the best option choice.

5. (b)

Cowardly is an adjective, so it cannot be used after the verb ***fight***.

After a verb we apply an adverb (except linking verb).

The adverb of ***cowardly*** is in a ***cowardly manner***.

6. (d)

We must know that, Hardly and Scarcely give negative sense.

To get good marks you have to work hard not hardly.

According to the above mentioned explanation options (a), (b) and (c) are proved to be wrong.

Thus option (d) is the best option choice.

7. (b)

The Rule says :

> ***Two much + uncountable noun + much too + adjective***

According to the above mentioned rules options (a), (c) and (d) are proved to be wrong.

Thus option (b) is the best option.

8. (c)

If a sentence begins with a negative adverb, we apply inversion.

Inversion means putting helping verb before the subject.

According to above mentioned rule, options (a), (b) and (d) are proved to be wrong.

Thus option (c) is the best option choice.

9. (c)

After seem we use an adjective ***sad*** is an adjective and ***sadly*** is an adverb, so the correct phrase is ***seems to be sad***.

Thus we can say that options (a), (b) and (d) are proved to be wrong.

10. (b)

Enough when as an adverb, it is placed after an adjective or adverb.

Thus options (a), (c) and (d) are wrong.

Thus option (b) is the best option choice.

CHAPTER 7

Conditional Sentences

Conditional Sentences

A conditional sentence is a statement that is used to discuss known factor or hypothetical situation and its result. A complete conditional sentence contains a conditional clause, also called if clause, and a result clause.

[If clause, + result clause]

EXAMPLES —

1. If you come to me, I will teach you.

   or

   I will teach you if you come to me.

2. If he had gone to Meerut, he would have seen that beautiful Church.

   or

   Had he gone to Meerut, he would have seen that beautiful Church.

NOTE

1. If clause is also called subordinate clause.	2. Result clause is also called main clause.

Classification of Conditional Sentences

There are four types of conditional sentences.

1. Type zero conditional sentences.
2. Type first conditional sentences.
3. Type second conditional sentences.
4. Type third conditional sentences.

1. Type Zero Conditional Sentences

Zero conditional sentences are used to express.

1. General truths
2. Scientific facts.

Rule : 1

	Conditional Clause	Result Clause
If +	Simple Present Present Continuous Present Perfect Present Perfect Continuous	+ Simple Present

Examples –

1. ☒ **Wrong** : If you heat iron, it will expand.
   ☑ **Right** : If you heat iron, it expands.
2. ☒ **Wrong** : If something is vibrating faster than twenty times a second it will audible to us.
   ☑ **Right** : If something is vibrating faster than twenty times a second it is audible to us.
3. ☑ **Right** : If it has been raining for several hours, the soil becomes saturated.

> ***When*** or ***whenever*** can be used in place of ***'if'*** in ***zero conditional*** sentences.

2. Type First Conditional Sentences:

Type first conditional sentences are used to express situations that may take place in the future.

Rule : 2

	Conditional Clause	Result Clause
If +	Simple Present Present Continuous Present Perfect Present Perfect Continuous	+ Sub + Modal + V1 + Obj

Modals Can, May, Might, Will, Shall etc.

Examples –

1. ☒ **Wrong** : If you will hurry, you might catch the bus.
   ☑ **Right** : If you hurry, you might catch the bus.
2. ☑ **Right** : If you are studying a foreign language, you have to make good command of vocabulary and grammar.
3. ☑ **Right** : If you keep driving like that, you are going to have an accident.
4. ☑ **Right** : If you like those flowers, I will buy them for you.

3. Type Second Conditional Sentences:

Type second conditional sentences are used to indicate unreal present situations.

Rule : 3

	Conditional Clause	Result Clause
If +	Simple Past Continuous	+ Subject + Would, Could, Might + V_1 + Object

If +	were + object + subject + would, could, might, should + V1 + object

Examples –

1. ☒ **Wrong** : If I knew the answer, I will tell you.
   ☑ **Right** : If I knew the answer, I would tell you.
2. ☑ **Right** : If she had money, she could help you.
3. ☑ **Right** : If you lived on another planed, the condition for life might be very difficult.
4. ☑ **Right** : If the trains were running on time, more people might take the train.
5. ☒ **Wrong** : If I was a better swimmer, I would try for the swim team.
   ☑ **Right** : If I were a better swimmer, If would try for the swim team.
6. ☑ **Right** : If we were free, we would love to go with you.

4. Type Third Conditional Sentences

Type third conditional sentences are sued to describe unreal past situations.

Rule : 4

	Conditional Clause	Result Clause
If +	Past Perfect, Past Perfect Progressive	+ Subject + would have / could have + V_3 + Object

V_3 Past Participle

Examples –

1. ☒ **Wrong** : If I knew the answer, I would have told you.
   ☑ **Right** : If I had known the answer, I would have to you.
1. ☒ **Wrong** : If you had waited for an hour, I would come with you.
   ☑ **Right** : If you had waited for an hour, I would have come with you.
   OR
   Had you waited for an hour, I would have come with you.
2. ☑ **Right** : If you had been listening to me, you should have heard what I said.
3. ☑ **Right** : If the teacher had not made the test so hard, I could have passed.

Some Other Usages of Conditional Sentences

Conditional sentences can be used to express ***'wishes and dreams'***. ***Wishes*** express things that are not true or real, but the speaker wants them to be true.

Rules :

Wish Clause		Result Clause	
Wish / If / If only	+ Simple Past, Subject +	Would / Could / Might	+ V_1 + Object

☒ **Wrong** : I wish I have a car.
☑ **Right** : I wish I had a car.
Means, I do not have a car.
☑ **Right** : I wish I knew where to find a beautiful used car.

If Only

If only is used to emphasise a hypothetical situations. With past events it adds a sense of regret.

Examples –

1. If only I had enough time.
2. If only I had not drunk too much, this would not have happened to me.

Happen to:

Happen to is used to emphasise the chance of possibilities.

Examples –

- If you happen to see Sandra, could you ask her to call me?
- If you should happen to be passing, drop in for a cup of tea.

But for

But for can be used at the place of if not.

Example –

- But for your help, I would have been in trouble.
  or
  If you had not helped me, I would have been in trouble.

Should

Should is used to make the possibility of an event seem unlikely.

Example –

- If you should see Alex, could you ask him to write me a letter?

PRACTICE EXERCISE Conditional Sentences

Instructions : Out of the following four sentences, select the most suitable sentence with respect to grammar and usage.

1. (a) If you heat iron, it will expands.
(b) If you will heat iron, it will expends.
(c) If you heat iron, it expands.
(d) If you heat iron, it can expands.

2. (a) If you come to me, I will teach you.
(b) If you came to me, I will teach you.
(c) If you will come to, I will teach you.
(d) If you would come to me, I should teach you.

3. (a) Were she a bird, she will fly in the sky.
(b) If she was a bird, she would fly in the sky.
(c) Were she a bird , she would fly in the sky.
(d) Was she a bird, she may fly in the sky.

4. (a) If he went to Agra, he would have seen the Taj Mahal.
(b) Had he gone to Agra, he can see the Taj Mahal.
(c) Had he gone to Agra, he would have seen the Taj Mahal.
(d) If he had gone to Agra, he should see the Taj Mahal.

5. (a) If she goes to market, she would have bought a tube of toothpaste.
(b) She would have bought a tube of toothpaste if she goes to market.
(c) If she goes to market, she will buy a tube of toothpaste.
(d) If she goes to market, she could have bought a tube of toothpaste.

6. (a) Had he come on time, the teacher would teach him.
(b) Had he come on time, the teacher would have teach him.
(c) Had he come on time, the teacher would have taught him.
(d) If he would have come on time, the teacher would have taught him.

7. (a) If I had studied more I would have done well on the test.
(b) If I study more I would have done well on the test.
(c) Had I studied more, I did well on the test.
(d) I would have done well on the test, if I will study more.

8. (a) If we will run fast, we will find him there.
(b) We will find him there if we will run fast.
(c) If we run fast, we could have found him there.
(d) If we run fast, we will find him there.

9. (a) If you had come to the theatre with me yesterday, you would enjoy the party.
(b) If you had come to the theatre with me yesterday, you would have enjoyed the party.
(c) If you had come to the theatre with me yesterday, you can enjoy the party.
(d) You can enjoy the party if you had come to the theater with me yesterday.

10. (a) If I knew the way, I would tell you.
(b) If I knew the way, I will tell you.
(c) If I knew the way, if can tell you.
(d) I can tell you if I knew the way.

Answer Key Conditional Sentences

1. (c)	2. (a)	3. (c)	4. (c)	5. (c)	6. (c)
7. (a)	8. (d)	9. (b)	10. (d)		

Explanations > Conditional Sentences

1. **(c)**

In case of zero conditional sentence, both the clauses should be in present indefinite tense only.

According to the above explanation, options (a), (b) and (d) are wrong.

Thus, (c) is the best option choice.

2. **(a)**

The Rule says :

If + Present Indefinite + Future Indefinite

According to the above rule options (a), (c) and (d) are termed wrong.

Thus, (a) is the best option choice.

3. **(c)**

The Rule says :

Were + subject + object, subject + would + V_1 *+ object*

According to the above rule options (a), (b) and (d) are proved to be wrong.

Hence option (c) is the best option choice.

4. **(c)**

The Rule says :

Had + subject + V_3 *+ object + subject + would have +* V_3 *+ object*

According to the above rule option (a), (b) and (d) are proved to be wrong.

Thus, option (c) is the best option choice.

5. **(c)**

The Rule says :

If + present indefinite tense + future indefinite tense

According to the above rule options (a), (b) and (d) are wrong.

Thus, option (c) is the best option choice.

6. **(c)**

The Rule says :

Had + subject + V_3 *+ object + subject + would have +* V_3 *+ object*

According to the above rule options (a), (b) and (d) are wrong.

Thus, option (c) is the best option choice.

7. **(a)**

The Rule says :

If + subject + had + V_3 *+ object + subject + would have +* V_3 *+ object*

According to the above rule options (b), (c) and (d) are wrong.

Thus, option (a) is the best option choice.

8. **(d)**

The Rule says :

If + present indefinite + future indefinite

According to the above rule options (a), (b) and (c) are wrong.

Thus, option (d) is the best option choice.

9. **(b)**

The Rule says :

If + subject + had + V_3 *+ object + subject + would have +* V_3 *+ object*

According to the above rule options (a), (c) and (d) are wrong.

Thus, option (b) is the best option choice

10. **(a)**

If + past indefinite tense + subject + would + V_1 *+ object*

According to the above rule options (b), (c) and (d) are wrong.

Thus option (a) is the best of option choice.

CHAPTER 8

Inversion

Inversion

Inversion in grammar means putting helping verb before the subject. We usually use inversion when we move a negative adverb such as hardly, scarcely, no sooner etc. to the beginning of a sentence. We do this because we want to emphasise the meaning of the statement.

Moreover inversion is the reversal of the normal word order in a sentence or phrase. There are two types of inversion.

1. Sub verb inversion *[Verb + subject]*
2. Sub-auxiliary inversion *[Aux verb + subject + verb]*

Examples –

1. Sub verb inversion

   On the top of the hill stood a stunning woman.

2. Along the street came a strange procession.

Sub-Auxiliary inversion

Examples –

1. Never had I seen such a beautiful sight before.
2. Only then did I understand why the tragedy had happened.

Now let me discuss **'inversion'** *with examples.*

We use inversion in the following situations:

1. In questions
2. With *so, neither, nor*

Examples –

1. Do you love your wife?
2. Have our father and mother arrived?
3. May all your wishes come true!
5. She speaks English so does her daughter.
6. Neha does not smoke neither does her husband.

☒ **Wrong** : Pecola drinks wine so her husband.

☑ **Right** : Pecola drinks wine, so does her husband.

☒ **Wrong** : I can solve these sums, so you.

☑ **Right** : I can solve these sums, so can you.

☒ **Wrong** : He does not go to school nor / neither does his sister.

☑ **Right** : He does not go to school nor / neither do his sisters.

We use inversion after negative and restrictive expressions.

☒ **Wrong** : Under no circumstances we can cash the cheques.

☑ **Right** : Under no circumstances can we cash the cheques.

☒ **Wrong** : Not until he received her letter he fully understood the depth of her feelings.

☑ **Right** : Not until he received her letter did he fully understand the depth of her feelings.

Explanation:

When a sentence begins with ***'Not until'*** the subject and helping verb are not inverted in the first clause, in the second clause the subject and auxiliary are inverted.

'Not until' means not before a stated time or event.

Let us have some more examples:

Examples –

1. Not until my mother called me, did I stop worrying about her.
2. Not until the rain stopped, could we see the view of the ocean.

   [Not until + sub verb object + helping verb + sub + verb + ...]

Inversion of conditional sentences:

Examples –

Normal word order

1. If he had gone to Agra, he would have seen the Taj Mahal.

Inverted order

Had he gone to Agra, he would have seen the Taj Mahal.

Normal word order

2. If she were a bird, she would fly in the sky.

Inverted order

Were she a bird she would fly in the sky.

When an adverbial expression of place or direction comes at the beginning of a sentence (clause), we use inversion.

Examples –

1. Under a tree was lying one of the biggest men I had over seen.
2. Directly infront of them stood a beautiful girl.
3. Along the road came a strange man.

We use inversion if the statement begins with the following negative adverbs / phrases, such as

Hardly, scarcely, rarely, in no way, never, seldom, so sooner, at no time, not only, only then, not for, not since, etc.

☒ **Wrong** : Never he comes on time.
☑ **Right** : Never does he come on time.

☒ **Wrong** : Hardly the teacher had entered the classroom when all the students stood up.
☑ **Right** : Hardly had the teacher entered the classroom when all the students stood up.

☒ **Wrong** : No sooner I reach the station than it began to rain.
☑ **Right** : No sooner did I reach the station than it began to rain.

☒ **Wrong** : Not for one minute I imagine my father will come back.
☑ **Right** : Not for one minute do I imagine my father will come back.

☒ **Wrong** : Not since I was little , I have had so much fun.
☑ **Right** : Not since I was little have I had so much fun.

PRACTICE EXERCISE

Adverbs / Inversion /Conditional Sentences

1. I am happy with (a) them because they (b) make regularly payments. (c) No error (d)

2. Love me (a) tender, and (b) treat me nicely. (c) No error (d)

3. You will have to (a) word your reply (b) very carefully. (c) No error (d)

4. Everybody knows (a) that they behaved (b) dreadful yesterday. (c) No error (d)

5. She will be working lately (a) not only today but also (b) every other day this week. (c) No error (d)

6. I love you, (a) but I am not in love with you (b) because you do not behave friendly. (c) No error (d)

7. Even after one year (a) of lessons Mona (b) plays the piano bad. (c) No error (d)

8. They walk (a) quick to (b) catch the train. (c) No error (d)

9. The driver (a) stopped (b) the bus abruptly. (c) No error (d)

10. During autumn colourful leaves (a) can be seen falling (b) much gently from trees. (c) No error (d)

11. If you heat (a) iron, (b) it will expand. (c) No error (d)

12. Seldom he (a) go to Church (b) on Sundays. (c) No error (d)

13. He does not (a) smoke nor (b) his wife. (c) No error (d)

14. The ship appeared (a) sudden (b) out of the fog. (c) No error (d)

15. If you look very carefully (a) the scratches on the car (b) are clearly visible. (c) No error (d)

16. From a distance, the hill appears (a) to rise steeply, but it (b) does not look steeply. (c) No error (d)

17. The forest smells most aromatically (a) when the wind (b) is from the south. (c) No error (d)

18. I feel sadly (a) about losing so much money (b) on the lottery. (c) No error (d)

19. One of my friends (a) is enough kind (b) to help you. (c) No error (d)

20. I have not (a) completed my (b) homework yet. (c) No error (d)

21. Morphology has been (a) defined the (b) study of word formation. (c) No error (d)

22. No sooner she had reached (a) the station than (b) it began to snow. (c) No error (d)

23. I seldom or ever (a) go there (b) to meet him. (c) No error (d)

24. I did my work (a) as better as (b) I could do. (c) No error (d)

25. You are advised (a) to keep your (b) books orderly. (c) No error (d)

26. Chirag Jainth is enough intelligent (a) to solve this question (b) in five minutes (c) No error (d)

27. Mother was beholding (a) her new born baby (b) when Clara entered the room (c) No error (d)

28. Everybody knows (a) that she is a (b) real good swimmer. (c) No error (d)

29. The new student (a) speaks French (b) poorly. (c) No error (d)

30. States have changed the way they finance education (a) allocating funds in extreme specific manners (b) rather than in large chunks. (c) No error (d)

31. Highway construction involves (a) a great deal of planning (b) as it can be quite unpredictably. (c) No error (d)

32. One of my (a) friends behaves (b) friendly (c) No error (d)

33. Few people will ever (a) remember too many (b) details from their early childhood. (c) No error (d)

34. I was very astonished (a) / to see him shouting (b) / at that woman. (c) / No error (d)

35. Never she had (a) / seen such a (b) / beautiful sight before. (c) / No error (d)

36. Not until I saw John (a) / with my own eyes (b) / did I really believe he was safe. (c) / No error (d)

37. Had he gone (a) / to Agra, he (b) / would see the Taj Mahal. (c) / No error (d)

38. Alex went (a) / to the club so (b) / his brothers. (c) / No error (d)

39. If I was you (a) / I would not (b) / do it. (c) / No error (d)

40. There on the tiny stoop (a) / sat Pecola (b) / in a light red sweater (c) / No error (d)

41. If I had studied (a) / more, I would have (b) / done good on the test. (c) / No error (d)

42. If you had come to the theatre (a) / with me yesterday (b) / you would enjoy the party. (c) / No error (d)

43. My brother often (a) / spends his free time in (b) / playing his guitar (c) / No error (d)

44. If we run (a) / fastly, we will (b) / find him there. (c) / No error (d)

45. I am too (a) / happy to (b) / see you smiling. (c) / No error (d)

46. If I said (a) / such a thing (b) / I should be ashamed (c) / No error (d)

47. Sandra works hardly (a) because she wants to get (b) good marks this year. (c) No error (d)

48. Have you seen (a) him reading (b) a detective novel late? (c) No error (d)

49. They stopped (a) briefly to fill (b) the car with petrol. (c) No error (d)

50. By looking close at DNA markers (a) scientist may have found traces of (b) the first African hunter-gatherers to migrate to other continents. (c) No error (d)

Answer Key > Adverb / Inversion / Conditional Sentences

1. (c)	2. (b)	3. (d)	4. (c)	5. (a)	6. (c)	7. (c)	8. (b)	9. (d)
10. (c)	11. (c)	12. (a)	13. (b)	14. (b)	15. (d)	16. (c)	17. (a)	18. (a)
19. (b)	20. (d)	21. (b)	22. (a)	23. (a)	24. (b)	25. (c)	26. (a)	27. (b)
28. (c)	29. (d)	30. (b)	31. (c)	32. (c)	33. (d)	34. (a)	35. (a)	36. (d)
37. (c)	38. (b)	39. (a)	40. (d)	41. (c)	42. (c)	43. (d)	44. (b)	45. (a)
46. (d)	47. (a)	48. (c)	49. (d)	50. (a)				

Explanations > Adverb / Inversion / Conditional Sentences

1. **(c)**
Replace *make regularly payments* by make regular payments.

2. **(b)**
Replace *tender, and* by tenderly, and.

3. **(d)**
No error.

4. **(c)**
Replace *dreadful yesterday* by dreadfully yesterday.

5. **(a)**
Replace *she will be working lately* by she will be working late.

6. **(c)**
Replace *because you do not behave friendly* by because you do not behave a friendly manner.

7. **(c)**
Replace *plays the piano bad* by plays the piano badly.

8. **(b)**
Replace *quick to* by quickly to.

9. **(d)**
No error.

10. **(c)**
Replace *much gently from trees* by very gently from trees.

11. **(c)**
Replace *it will expand* by it expands.

12. **(a)**
Replace *seldom he* by seldom does he.

13. **(b)**
Replace *smoke nor* by smoke nor does.

14. **(b)**
Replace *sudden* by suddenly.

15. **(d)**
No error.

16. (c)
Replace *doe not look steeply* by does not look steep.

17. (a)
Replace the *forest smells most aromatically* by the forest smells the most aromatic.

18. (a)
Replace *I feel sadly* by I feel sad.

19. (b)
Replace is enough kind by is kind enough.

20. (d)
No error.

21. (b)
Replace defined the by defined as the.

22. (a)
Replace no sooner she had reached by no sooner had she reached.

23. (a)
Replace I seldom or ever by I seldom or never.

24. (b)
Replace as betters by as well as.

25. (c)
Replace books orderly by in a orderly manner.

26. (a)
Replace *Chirag Jainth is enough* intelligent by Chairag Jainth is intelligent enough.

27. (b)
Replace *her new born baby* by her newly born baby.

28. (c)
Replace *real good swimmer* by realy good swimmer.

29. (d)
No error.

30. (b)
Replace *allocating funds in extreme specific manners* by allocating funds in extremely specific manners.

31. (c)
Replace *as it can be quite unpredictably* by as it can be quite unpredictable.

32. (c)
Replace *friendly* by in a friendly manner.

33. (d)
No error.

34. (d)
Replace *I was very astonished* by I was much astonished.

35. (a)
Replace *never she had* by never had she.

36. (d)
No error.

37. (c)
Replace *would see the Taj Mahal* by would have seen the Taj Mahal.

38. (b)
Replace *to the club*, so by to the club, so did.

39. (a)
Replace *If I was you* by If I were you.

40. (d)
No error.

41. (c)
Replace *done good on the test* by done well on the test.

42. (c)
Replace *you would enjoy the party* by you would have enjoyed the party.

43. (d)
No error.

44. (b)
Replace *fastly, we will* by fast, we will.

45. (a)
Replace *I am too* by I am very.

46. (d)
No error.

47. (a)
Replace *Sandra works hardly* by Sandra works hard.

48. (c)
Replace *a detective novel late* by a detective novel lately.

49. (d)
No error.

50. (a)
Replace *by looking close at DNA markers* by looking closely at DNA markers.

■

CHAPTER 9

Question Tag

Question Tag

A short question which is added at the end of a statement. We use question tag to

1. Confirm whether something is true or no.
2. Get response from the person, we are speaking to.

Examples –

Sohan writes a book, doesn't he?

They helped the poor, didn't they?

How to form a question tag?

A question tag is formed with the auxiliary or modal verb form, and the appropriate subject pronoun. (i.e. ***I, we, you, he, she, it , they,*** etc.)

Points to Remember

- Always use pronoun in question tag
- Always use contract form (doesn't, won't, haven't, etc.)
- Add a negative question tag with positive statement.
- Add a positive question tag with negative statement.
- Use a comma before the questions tag.

Guidelines

1. If the verb is in the present simple, we form the question tag with ***do / does***.

   Examples –

   - Nidhi composes a song, doesn't she?
   - They do not go home, do they?
   - Alison and Cathy help the poor, don't they?

2. If the verb is in the simple past we form the question tag with did.

   Examples –

   - Rochester wrote an essay, didn't he?
   - Sita went to school, didn't she?
   - They did not help us, did they?
   - Mother used to take tea, didn't she?

3. If the statement contains a word which is negative in meaning, the question tag should be positive.

   i.e. ***hardly, scarcely rarely, seldom, nothing, nobody, no one, never, few, little, etc.***

EXAMPLES –

- She hardly comes on time, does she?
- Barking dogs seldom bite, do they?
- The baby needs little milk, does it?
- I bought few books, did I?

4. If a statement begins with ***someone, somebody, everyone, everybody, no one, nobody,*** the pronoun of the question tag should be ***'they'***.

EXAMPLES –

- Somebody is coming, aren't they?
- Nobody can solve these sums, can they?
- Everybody was waiting for her, weren't they?
- No one has eaten food yet, have they?

The contract form of ***am not*** is ***aren't***

- I am reading a book, aren't I?
- I am not reading a book, am I?

Imperative Sentences

- If an imperative sentence is ***positive***, we can add ***will you?*** or ***won't you?,*** and ***sometimes could you?*** but in case of negative we add ***will you?***
  1. Open the door, ***will you***?

     OR

     Open the door, ***won't you***?

     OR

     Open the door, ***could you***?
  2. Water the plants, ***will you***?

     OR

     Water the plants, ***won't you***?
  3. Do not touch that wire, will you?

- The question tag of ***'let us'*** is always ***'shall we'***?
  1. Let us go home, shall we?
  2. Let us play hockey, shall we?
  3. Let us not go there, shall we ? → ***[Rarely do we use this type of statement]***

- A collective noun takes a singular verb and a singular pronoun ***(it)***
  1. This team is playing well, isn't it?
  2. The jury has given its decision, hasn't it?

- When we talk about the individual members of the collective noun,
  1. Use plural verb.
  2. Use plural pronoun ***'They'***

  EXAMPLES –

  1. The team are sitting, aren't they?
  2. The audience have taken their seats, haven't they?

- Demonstrative pronoun this and that change into *'it'* and these and those change it *'they'* but there is no change into *'there'*.
  1. This is a good pen, isn't it?
  2. That is a pretty flower, wasn't it?
  3. There is not water in this glass, is there?

Now read each example carefully with a view of making the better understanding of the questions?

1. She cut vegetables, didn't she?
2. He cuts trees, doesn't he?
3. We have a car, don't we?
4. We have got a car, haven't we?
5. They have eaten food, haven't they?
6. They have food, don't they?
7. We can solve these sums, can't we?
8. Something is burning, isn't it?
9. We will not compose a song, shall we?

NOTE : In interrogative statement we use shall with *I, we*.

10. This is a good book, isn't it?
11. That was my car, wasn't it?
12. There are forty students in my class, aren't there?
13. The jury has taken its decision, hasn't it?
14. Let us go home, shall we?

PRACTICE EXERCISE Question Tag

Add question tag to the following statements

1. Meera is going home, ...
2. We are listening to the music, ...
3. Something is smouldering, ...
4. Ram and Mohan help the poor, ...
5. He is coming tonight, ...
6. You wouldn't report me, ...
7. There are fifty people in that room, ...
8. Please help this man, ...
9. I am composing an elegy, ...
10. Rahul has a car, ...
11. Sheela has got a scooter, ...

12. I am not going to market, ...
13. Neha went to the school to meet her son, ...
14. We will not count on you, ...
15. He seldom comes to meet me, ...
16. Nobody lives in this house, ...
17. She needs little water, ...
18. We had a cup of tea, ...
19. She will come tomorrow, ...
20. Have some more tea, ...
21. Let us sit there, ...
22. It hardly snows here, ...
23. The mob killed the officer, ...
24. They used to smoke, ...
25. I am tired, ...
26. There is no work to do, ...
27. We want to help them, ...
28. Meera has written several novels, ...
29. Ansh has painted the box green, ...
30. Mother seems to be cooking food, ...
31. God helps us, ...
32. She does not live in Delhi, ...
33. My friends are waiting for you, ...
34. Those were my books, ...
35. Nothing is burning, ...

Answer Key

1.	isn't she?	8.	will you? won't your?	15.	does he?	22.	does it?	29.	hasn't he?
2.	aren't we?	9.	aren't I?	16.	do they?	23.	didn't it?	30.	doesn't she?
3.	isn't it?	10.	doesn't he?	17.	does she?	24.	didn't they?	31.	doesn't he?
4.	don't they?	11.	hasn't she?	18.	didn't we?	25.	aren't I?	32.	does she?
5.	isn't he?	12.	am I?	19.	won't she?	26.	is there?	33.	aren't they?
6.	would you?	13.	didn't she?	20.	will you?	27.	don't we?	34.	weren't they?
7.	aren't there?	14.	shall we?	21.	shall we?	28.	hasn't she?	35.	is it?

CHAPTER 10

Subject Verb Agreement

Subject Verb Agreement

Subject Verb Agreement is the most basic concept in the English Language. It means to apply verb according to subject. Sub-Verb Agreement includes two important rules.

1. **Singular subject takes singular verb.**
2. **Plural subject takes plural verb.**

Key points

1. Base form is always plural, so add ***s/es*** to make it singular.

Plural Verb	Singular Verb
go read ameliorate wash cry	goes reads ameliorates washes cries

2. Be-verbs (i.e. ***is / am / are, was, were***) must agree with the subject.
3. Modal verbs do not have any agreement with the subject.
4. A noun is in singular form, so add ***s / es*** to make it plural.

Singular Noun	Plural Noun
Book Teacher Pen	Books Teachers Pens

Singular helping Verb	Plural helping Verb
is am was has does	are were have do

The rules of ***Subject Verb Agreement*** are applied on the following tenses.

1. Simple present
2. Present progressive
3. Present perfect
4. Present perfect progressive
5. Past progressive

NOTE

Please do read ***'the tenses'***, ***'nouns'***, ***'pronouns'*** and ***'conjunctions'*** thoroughly to have through knowledge of ***'Subject Verb Agreement'***.

Rules Regarding Subject Verb Agreement

Rule : 1 A singular subject takes a singular verb.

Examples –

1. Rochester comes on time to attend the class.
2. Belinda composes a sonnet.
3. Ansh plays cricket.
4. It has been raining since morning.
5. I was writing an essay when they entered the room.

A plural subject takes a plural verb.

Examples –

1. They help the poor.
2. We were waiting for you.
3. You are watching a movie.
4. They do not live in this town.

More about Sub-Verb Agreement

Compound Subject: Subjects linked equally with ***'and'*** (called ***compound subjects***) they have plural verb.

- Chirag and Ansh have gone to market with a view to purchasing a few articles of furniture.
- What he says and what he does are never consistent.

Practice Questions :

☒ **Ram and Mohan seems to be happy today.**
☑ **Ram and Mohan seem to be happy today.**

☒ **What she speaks and what she does is never consistent.**
☑ **What she speaks and what she does are never consistent.**

NOTE

Be sure that you must have a true ***compound subject*** before making the ***verb plural***.

Rule : 2 Agreement with Lost Subjects.

Nouns or ***pronouns*** introduced after the subject with phrasal prepositions such as - ***as well as, with, along with, together with, and not, in addition to, of, rather than, not to mention, including, accompanied by***, are not part of the subject. They are ***subordinate to, not equal to***, the subject.

Subject-1 +

along with
as well as
with
together with
and not
not to mention
of
accompanied by
rather than
in addition to
of

\+ Subject-2 + Verb

☒ **Wrong** : The vice present, together with three managers from Canada are responsible for the project.

☑ **Right** : The vice president together with three managers from Canada is responsible for the project.
Here, our real subject is ***'The vice president'*** which is singular.

☒ **Wrong** : The cost of all the repairs, we needed to make were more than we could afford.

☑ **Right** : The cost of all the repairs, we needed to make ***was*** more than we could afford.

☒ **Wrong** : A group of students want to meet the principal regarding this case.

☑ **Right** : A group of students ***wants*** to meet the principal regarding this case.

☒ **Wrong** : Rochester and not his relatives are coming to meet me.

☑ **Right** : Rochester and not his relatives ***is*** coming to meet me.

☒ **Wrong** : A bouquet of red roses lend colour and fragrance to the room.

☑ **Right** : A bouquet of red roses lends colour and fragrance to the room.

☒ **Wrong** : Developments in neuroscience is challenging long held beliefs about how people process information during stressful situations.

☑ **Right** : Developments in neuroscience ***are*** challenging long held beliefs about how people process information during stressful situations.

Rule : 3 **In case of *neither - nor*, *either - or*, *or*, *not only* - *but also*,** make the verb agree with the closest subject.

☒ **Wrong** : Either Ram or Mohan have taken your book.

☑ **Right** : Either Ram or Mohan has taken your book.

☒ **Wrong** : Neither Meera nor her friends is used to taking tea.

☑ **Right** : Neither Meera nor her friends are used to taking tea.

☒ **Wrong** : Not only bicycling but also sail boating are being planned for the company's memorial day picnic.

☑ **Right** : Not only bicycling but also sail boating is being planned for the company's memorial day picnic.

Rule : 4 **You (pronoun)** may be ***singular*** or ***plural*** in meaning, but it always requires a ***plural verb***.

Examples –

1. You are going home.
2. You go to Church on Sundays.

Rule : 5 When two or more nouns are joined with ***'and'*** to form a single unit, idea or result, the verb must be ***singular***.

☒ **Wrong** : Bread and butter make a delicious breakfast.
☑ **Right** : Bread and butter ***makes*** a delicious breakfast.

☒ **Wrong** : The horse and the cart are running on the road.
☑ **Right** : The horse and the cart ***is*** running on the road.

☒ **Wrong** : Slow and steady win the race.
☑ **Right** : Slow and steady ***wins*** the race.

Rule : 6 **Indefinite pronouns such as**

Somebody, anybody, each, everyone, nobody, someone, anyone, either, everything, no one, something, anything, everybody, neither, nothing, etc. Take Singular verbs, even though they may seem to have plural meanings.

☒ **Wrong** : Everyone have an equal chance of winning the lottery.
☑ **Right** : Everyone ***has*** an equal chance of winning the lottery.

☒ **Wrong** : Everything alive on our planet are composed of cells.
☑ **Right** : Everything alive on our plant ***is*** composed of cells.

☒ **Wrong** : Nobody come on time to attend the seminar.
☑ **Right** : Nobody ***comes*** on time to attend the seminar.

Rule : 7 **One of + plural noun + singular verb**

☒ **Wrong** : One of my friends live in America.
☑ **Right** : One of my friends lives in America.

Rule : 8 **The number of + plural noun + singular verb**

☒ **Wrong** : The number of students are increasing rapidly.
☑ **Right** : The number of students ***is*** increasing rapidly.

Rule : 9 **A number of + plural noun + plural verb**

☒ **Wrong** : A number of students is increasing rapidly.
☑ **Right** : A number of students ***are*** increasing rapidly.

Rule : 10 When a singular noun represents two or more posts; use singular verbs.

☒ **Wrong** : The principal and class teacher often come late to teach as.
☑ **Right** : The principal and class teacher often comes late to teach us.

> **NOTE**
>
> Here we are talking about *one person*.
> The principal and the class teacher often come late to teach us.
>
> This sentence is correct because *'The'* definite article has been used twice, which means we are talking about two different people.

Rule : 11 Fraction / Percentage / Portions / Distance, etc.

Another special case is the use of fractions, percentage in or other words that mean part of a whole.

Choose a plural verb when the sentence implies a plural noun, and choose a singular verb when the sentence implies a singular noun.

☒ **Wrong** : One half of the mailing list have been deleted.
☑ **Right** : One half of the mailing list ***has*** been deleted.
☒ **Wrong** : Forty percent of this apple are rotten.
☑ **Right** : Forty percent of this apple ***is*** rotten

☒ **Wrong** : One fourth of the booklets has missing pages.
☑ **Right** : One fourth of the booklets ***have*** missing pages.

☒ **Wrong** : Ten kilometers are not a long distance
☑ **Right** : Ten kilometers ***is*** not a long distance.

Rule : 12 Usage of 'There' as a subject

Actually, there construction is called existential. Existential sentences are used to point out the existence of something.

Examples –

1. There was a king.
2. There are seventy students in my class.
3. There is no water in this glass.
4. There appears to be no solution.
5. There was a bright light shining in the trees.

> **NOTE**
>
> In case of *'There'* the verb is used according to the object.

Rule : 13 Each and Every

When the modifier each or every is used to modify a compound subject, the verb must be singular.

☒ **Wrong** : Each boy and girl have completed her homework.
☑ **Right** : Each boy and girl has completed her homework.

☒ **Wrong** : It seems that every town, country, and village have a number of small shops.
☑ **Right** : It seems that every town, country, and village has a number of small shops.

☒ **Wrong** : Each of the passengers wish to get off here.
☑ **Right** : Each of the passengers wishes to get off here.

Rule : 14 Inverted Sentence Order

Usually in a sentence the subject precedes the ***verb***, but sometimes this order is ***reversed***. It means the ***verb*** comes before the ***subject***.

☒ **Wrong** : Through this bridge passes thousands of people everyday.
☑ **Right** : Through this bridge pass thousands of people everyday.

☒ **Wrong** : Here is the blue buttons to sew on my trousers.
☑ **Right** : Here are the blue buttons to sew on my trousers.

Rule : 15 Relative Pronouns as Subjects

A relative pronoun such as who, which, or that has no number within itself, so the verb is used according to the pronoun or noun that precedes it.

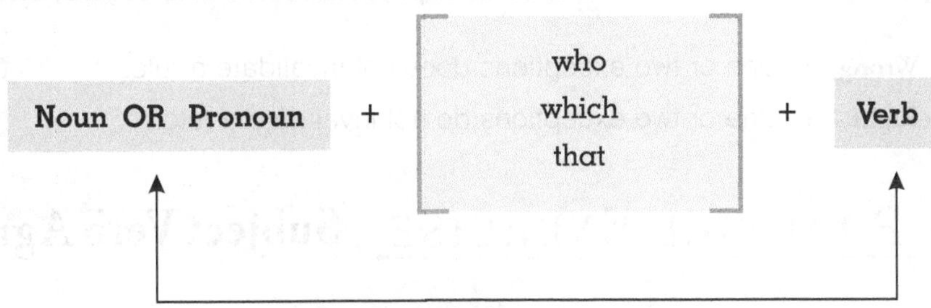

☒ **Wrong** : It is I who is going home with them.
☑ **Right** : It is I who am going home with them.

☒ **Wrong** : My mother is one of the reporters who covers the City Hall.
☑ **Right** : My mother is one of the reporters who cover the City Hall.

☒ **Wrong** : Candida is one of those people who gives cheap gifts.
☑ **Right** : Candida is one of those people who give cheap gifts.

Rule : 16 But in case of the only, the verb is singular.

☒ **Wrong** : She is the only one of those commuters that take morning train.
☑ **Right** : She is the only one of those commuters that takes morning train.

Rule : 17 In case of more than one, use singular sub and singular verb

More than + singular noun + singular verb

☒ **Wrong** : More than one person were killed in that blast.
☑ **Right** : More than one person was killed in that blast.

In case of More + Plural noun + than one + plural verb

☒ **Wrong** : More people than one was killed in that blast.
☑ **Right** : More people than one were killed in that blast.

Rule : 18 Names of the books, magazines, companies (whether singular or plural) take singular verb.

☒ **Wrong** : Midnight's children are a good novel.
☑ **Right** : Midnight's children is a good novel.

Rule : 19 In case of an uncountable noun, a gerund or a noun clause, the verb should be singular.

☒ **Wrong** : Milk contain several useful vitamins.
☑ **Right** : Milk contains several useful vitamins.

☒ **Wrong** : The reading of these books seem to be essential for you and her.
☑ **Right** : The reading of these books seems to be essential for you and her.

☒ **Wrong** : How animals interact with each other interest their trainers.
☑ **Right** : How animals interact with each other interests their trainers.

Rule : 20 A plural verb is used in the case of expressions like one or two, one or both one or more.

☒ **Wrong** : One or two exceptions does not invalidate a rule.
☑ **Right** : One or two exceptions do not invalidate a rule.

PRACTICE EXERCISE Subject Verb Agreement

Instructions : Out of the following four sentences, select the most suitable sentence with respect to grammar and usage.

1. (a) A team of graduate students are going to Agra to see the Taj Mahal.
(b) A team of graduate student are going to Agra to see the Taj Mahal.
(c) A team of graduate students were going to Agra to see the Taj Mahal.
(d) A team of graduate students is going to Agra to see the Taj Mahal.

2. (a) Each of the dogs in the show require a special type of diet.
(b) Each of the dogs in the show requires a special type of diet.
(c) Each of the dog in the show requires a special type of diet.
(d) Each of the dogs in the show have required a special type of diet.

3. (a) Adding extra passengers cause a boat to settle more deeply in water.
(b) Adding extra passengers causes a boat to settle more deeply in water.
(c) Adding extra passengers have caused a boat to settle more deeply in water.
(d) Adding extra passengers are causing a boat to settle more deeply in water.

4. (a) One of my relative lives in Delhi.
(b) One of my relatives live in Delhi.
(c) One of my relatives were living in Delhi.
(d) One of my relatives lives in Delhi.

5. (a) Some of the money were stolen from ,my wallet.
(b) Some of the money have been stolen from my wallet.
(c) Some of the money are stolen from my wallet.
(d) Some of the money was stolen from my wallet.

6. (a) The people living the next door play loud music.
   (b) The people living the next door plays loud music.
   (c) The people living the next door is playing loud music.
   (d) The people living the next door has played loud music.

7. (a) Neither Ram nor his friends has composed a sonnet.
   (b) Neither Ram nor his friends have composed a sonnet.
   (c) Neither Ram nor his friends is composing a sonnet.
   (d) Neither Ram nor his friends composes a sonnet.

8. (a) The little girl in the light blue jumper and matching shoes are crying.
   (b) The little girl in the light blue jumper and matching shoes is crying.
   (c) The little girl in the light blue jumper and matching shoes seem to be crying.
   (d) The little girl in the light blue jumper and matching shoes have been crying.

9. (a) The number of students were increasing rapidly.
   (b) A number of students was increasing rapidly.
   (c) The number of students was increasing rapidly.
   (d) The number of students have been increasing rapidly.

10. (a) One of the biggest problems are the amount of money that they have to spend on tent.
    (b) One of the biggest problems is the amount of money that they have to spend on tent.
    (c) One of the biggest problem is the amount of money that they has to spend on tent.
    (d) One of the biggest problems were the amount of money that they have to spend on tent.

Answer Key ➤ Subject Verb Agreement

1. (d)	2. (b)	3. (b)	4. (d)	5. (d)	6. (a)
7. (b)	8. (b)	9. (c)	10. (b)		

Explanations ➤ Subject Verb Agreement

1. (d)

Option (a) is wrong because the verb should be singular.

'A team of graduate student' is a wrong phrase so option (b) is wrong.

Option (c) is wrong because it also has verb error.

Thus option (d) is the best option choice.

2. (b)

The Rule says :

Each of + Plural noun + Singular verb

According to the above mentioned rule options (a), (c) and (d) are wrong.

Thus option (b) is the best option choice.

3. (b)

Gerund *[V_1 + ing]* used as a noun takes singular verb.

Example : *Reading novels gives me too much pleasure.*

After making the close study of the above mentioned rule we come to know that options (a), (c) and are wrong.

Thus option (b) is the best option choice.

4. (d)

The Rule says :

One of + Plural noun + Singular verb

According to above mentioned rule options (a), (b) and (c) are wrong. So option (d) is the best option choice.

5. (d)

The Rule says :

Some of + Uncountable noun + Singular verb

If we read the options (a), (b) and (c), we will come to that they are wrong because money is an uncountable noun, so it requires a singular verb.

Thus option (d) is the best option choice.

6. (a)

Option (b) and (c) are wrong because the people is a plural noun.

Option (d) is also wrong because with the people we cannot use has.

Thus option (a) is the best option choice.

7. (b)

In case of ***[neither .... nor]*** the verb is applied according to the nearest subject. Here our subject is ***students***, so the verb should be used according to ***students*** i.e. have.

According to the above mentioned explanation options (a), (c) and (d) are wrong.

Thus option (b) is the best option choice.

8. (b)

In these sentences our subject is ***The little girl***, so we have to apply singular verb i.e. ***is***.

According to above mentioned explanation options (a), (c) and are wrong.

Thus option (b) is the best option choice.

9. (c)

The Rule says :

The number of + Plural noun + Singular verb

According to rule mentioned above options (a), (b) and (d) are wrong.

Thus option (c) is the best option choice.

10. (b)

The Rule says :

One of + Plural noun + Singular verb

According to above mentioned rule options (a), (c) and are wrong.

Thus option (b) is the best option.

CHAPTER 11

Non Finite Verbs

Finite Verb

A verb is said to be finite when it shows tense, person and number. We can say a finite verb changes according to the subject or the tense.

Examples —

1. She ***goes*** to school.
2. They ***go*** to school.
3. They went to school yesterday.

Here we see **(1, 2)** that the verbs ***goes*** and ***go*** are changing their forms according the subjects (i.e. She, They respectively). In the third sentence ***go*** becomes went because of past tense.

NOTE

- Most verbs can be ***finite*** or ***Non-finite***.
- Whether they are ***finite*** or ***non-finite*** depend on the way, we use them.

Non-finite Verb

A verb is said to 'non-finite verb' when it does not change according to the subject or the tense.

Examples —

1. He goes ***to play*** hockey.
2. They go ***to play*** hockey.
3. She went ***to play*** hockey.

   In the above sentences ***'To play'*** is a non-finite verb, because it does not change according to the subject or the tense.

Classification of Non-finite Verbs

Three types:

1. Infinitives
2. Gerunds
3. Participles

Infinitives

The base form of the verb is known as an infinitive, such as ***to go***, ***to write***, ***to sing***, ***to play*** etc.
Infinitive can be divided into two categories.

1. **Bare Infinitive :** Base form such as ***read, write, sing, play,*** etc.
2. **Full infinitive :** to + base form, such as ***to read, to write, to play*** etc.

Moreover an infinitive can be

1. **Simple :** *(To + V_1)*

   Ex. : He wants to write an essay.

2. **Progressive :** *(To + be + V_1 + ing)*

   Ex. : She seems ***to be*** composing a song.

   OR

   It seems that she is composing a song.

3. **Perfect :** *(To + have + V_3)*

   Ex. : He seems ***to have*** eaten food.

   OR

   It seems that he has eaten food.

Error in the usage of Infinitives:

Bare infinitive [V_1] also called base form, principal verb, main verb, root verb, lexical verb etc.

Rule : 1 **Had better** takes bare infinitive.

☒ **Wrong** : You had better to write in pencil.
☑ **Right** : You had better write in pencil.

☒ **Wrong** : Sandra had better to take this medicine regularly.
☑ **Right** : Sandra had better take this medicine regularly.

Rule : 2 We use bare infinitive after

1. Modals, also known as modal auxiliaries verbs such as will, shall, can, could, may, might, ought to, etc.

   Ex.: • I could write a novel.
   • We should help the poor and the needy.

2. After would rather

☒ **Wrong** : I would rather to stay at home than go out tonight.
☑ **Right** : I would rather stay at home than go out tonight.

Rule : 3 ***'Let'*** also takes bare infinitive.

☒ **Wrong** : Let you and me to compose a sonnet.
☑ **Right** : Let you and me compose a sonnet.

Rule : 4 ***Need*** and ***Dare*** can be used as modal auxiliaries and as main verbs.

Example –

1. I need to write an essay.
   ↓
   (main verb)

2. I need not write an essay
   ↓
   (modal auxiliary)

REMEMBER

Use ***(To + V_1)*** when ***need*** and ***dare*** used as main verbs

Need / Dare + Not

Helping verb + Need/ Dare (main verbs)

3. I do not need to write an essay.
↓
(main verb)

☒ **Wrong** : She does not need go there to meet that stupid man.
☑ **Right** : She does not need to go there to meet that stupid man.

Rule : 5 Special usage of ***need / In need of***

1. When the subject of a sentence is living being, the verb ***'need'*** is followed by ***(To + V1)***

   Example –
   - Peter ***needs*** to get a haircut.
   - My girlfriend ***needs*** to learn new tricks.
   - You ***need*** to leave now.
2. When the subject of a sentence is a non living object, the verb need is followed by either a gerund ***[V1 + ing used as a noun] or [to + be + past participle (V3)]***

   Example –
   - The grass ***needs*** cutting.

     OR

     The grass ***needs to*** be cut.
   - This omnibus ***needs*** repairing.

     OR

     This omnibus ***needs to*** be repaired.

   NOTE **in the need of** can be used at the place of ***need***

   Sachin needs a haircut.
   OR Sachin is in the need of a haircut.

Rule : 6 We use bare infinitive with causative verbs such as make, made, will make, help, etc.

☒ **Wrong** : Rohan ***made*** the horse to run.
☑ **Right** : Rohan ***made*** the horse run.

☒ **Wrong** : Candida ***will make*** the children to laugh.
☑ **Right** : Candida ***will make*** the children laugh.

Gerunds

A gerund is formed by adding an ***'ing'*** form to the base form of a verb, and it acts like a noun, such as

1. Reading is a good habit.
2. Smoking is injurious to health.

Since a gerund is a kind of noun, it can work as a/an

1. Subject
2. Object
3. Complement
4. Object of preposition

EXAMPLES —

1. Reading is a good habit.
↓
Subject

2. I love reading books.
   ↓
   Object

3. May bad habit was sleeping in the morning.
   ↓
   Complement

4. She is used to smoking
   ↓ ↓
   Preposition Object

Errors Gerunds vs Infinitives

Rule : 1 Gerund and Infinitive should not be mixed in a sentence.

☒ **Wrong** : I like reading, writing, speaking and to compose poems in my spare time.
☑ **Right** : I like reading, writing, speaking and composing poems in my spare time.

Rule : 2 We do not use *(to + V)* infinitive after preposition.

☒ **Wrong** : I am fond of read novels.
☑ **Right** : I am fond of reading novels.
☒ **Wrong** : I am looking forward to meet you.
☑ **Right** : I am looking forward to meeting you.

Always use 'gerund' after the following prepositions

Look forward to, Key to, Be + used to, Accustomed to, Prone to, Addicted to, With a view to, Committed to, Object to	**+ V_1 + ing**

Let us have some more ***Examples***:
He is used to taking tea.
He came here with a view to meeting me.
OR
He is accustomed to taking tea.

NOTE

Usually we apply bare form after to

To + V_1 → infinitive	*To* + V_1 + *ing* ↓ Preposition

Now clear your doubt

How to decide *'to'* is an infinitive indicator or a preposition. Now to clear your doubt look at this simple tip:

- If you put a noun after 'to' and it makes sense, *'to'* is a preposition, and you can put an *'ing'* form (i.e. gerund) after *'to'*.

Key to + Success
↓ (makes sense) ↘ (noun)
(to) is a preposition

Now it is ok to say

There is no key to learning English without proper guidance and proper practice.

Rule : 3 Use 'gerund' after the following verbs:

admit, appreciate, avoid, complete, delay, discuss, enjoy, finish, miss, mind, insist on, imagine, postpone, prevent, risk, resume, resist, stop, consider, etc.

☒ **Wrong** : Would you ***mind*** my come to you at night.
☑ **Right** : Would you ***mind*** my coming to you at night?

☒ **Wrong** : The committee chairman will ***consider*** to add three items to the agenda.
☑ **Right** : The committee chairman will ***consider*** adding three items to the agenda.

NOW READ

- He ***stopped*** writing. [He was not writing when I saw him]
- He ***stopped*** to write. [He was doing something else and began to write.]

[It means, sometimes ***'ing'*** form and ***'To + V1'*** have different meanings.]

Rule : 4 Use infinitive ***(To + V1)*** after the following adjectives:

able, afraid, difficult, delighted, glad, sorry, good, happy ready, pleased, sad, possible, proud, motivated, important, eager, disappointed, determined, etc.

☒ **Wrong** : I was ***sorry*** hearing about your accident.
☑ **Right** : I was ***sorry*** to hear about your accident.

NOTE Sometimes ***(for + noun)*** comes between an adjective and an infinitive.

Examples –

- It is ***dangerous*** for everyone to smoke.
- It is very ***important*** for every student to remember the rules of English grammar to speak well.

Rule : 5 To indicate purpose, we use an infinitive ***(To + V1)***, not a gerund.

☒ **Wrong** : I am going to market for purchasing a tube of toothpaste.
☑ **Right** : I am going to market to purchase a tube of toothpaste.

☒ **Wrong** : Candida came for meeting me.
☑ **Right** : Candida came to meet me.

Rule : 6 Use infinitive after the following verbs:

Advise, agree, allow, ask, attempt, choose, pretend, prepare, tend, plan, want, need, seem, appear, intend, prove, wish, work, expect, beg, decide, hope, get, etc.

☒ **Wrong** : Do you care commenting on that beautiful woman?
☑ **Right** : Do you care to comment on that beautiful woman?

☒ **Wrong** : Please do not hesitate calling me if you have any problem.
☑ **Right** : Please do not hesitate to call me if you have any problem.

Participles

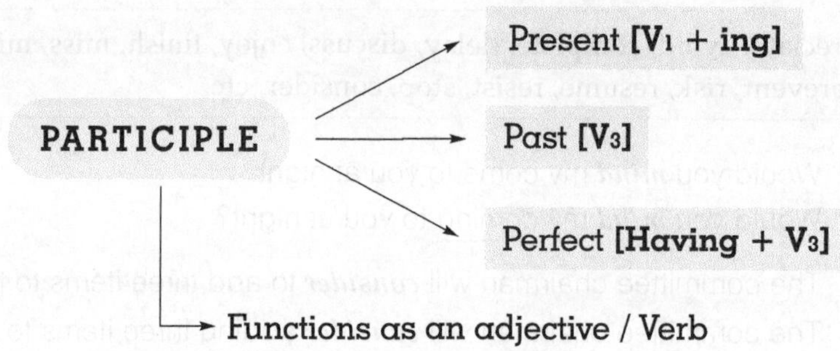

A participle functions as an adjective. It is also used to make continuous.

1. I saw him reading. *[Present Participle]*
2. I found my lost pen. *[Past Participle]*
3. Having written a letter, he posted it. *[Perfect participle]*

▣ PRESENT PARTICIPLE

Formation *[Base form + ing]*

Functions as an adjective.

Examples –

1. I saw Mona reading a novel.
2. Jumping up, he ran away.

Present participle is also used to join two actions, that take place simultaneously, as in ***Example (2)***.

▣ PAST PARTICIPLE

[V_3] such as wounded, lost, broken, etc.

Example –

1. I found my lost pen.
2. Where is your broken chair?

In the examples, given above, ***lost*** and ***broken*** have been used as adjectives.

▣ PERFECT PARTICIPLE *[Having + Past + Participle]*

Examples –

1. Having written a letter, he posted it.
2. Having completed his home work he went out to play hockey.

CHAPTER 12

Parallelism

Parallelism

Sentences frequently contain lists of two or more items. Such items in a sentence should be parallel; that is they should have the same grammatical structure.

Parallelism is the placement of equal ideas in words, phrases or clauses of similar type.

Parallel structure with Nouns

Example –

- I would like to purchase a book, (*Noun*) a pen (*Noun*) and a pencil (*Noun*)

  The above sentence is parallel.

  ☒ **Wrong** : The qualities of a good man are loyalty, honesty and being brave.

  ☑ **Right** : The qualities of a good man are loyalty, honesty and ***bravery***.

Being ***brave*** is a ***phrase*** not a **noun**.

Parallel structure with Adverbs

Example –

- Belinda walked ***slowly (Adv)*** and ***lovingly (Adv)*** towards me.

  ☒ **Wrong** : A well trained dog obeys willingly, complete and quickly.

  ☑ **Right** : A well trained dog obeys willingly, ***completely*** and quickly.

Parallel structure with adjectives

Example –

- One of my friends is ***intelligent (Adj)*** and ***honest (Adj)***.

Parallel structure with verbs

Example –

- They ***looted (verb)*** the house and ***took (verb)*** away the things.

  ☒ **Wrong** : The class analysed, discussed, and we raised questions about the new rules.

  ☑ **Right** : The class analysed, discussed, and questioned the new rules.

  ☒ **Wrong** : Sohan prepared the presentation on phonetics and it was delivered by him at the seminar.

  ☑ **Right** : Sohan prepared the presentation on phonetics and delivered it at the seminar.

Parallel structure with infinitives and gerunds

Infinitives and Gerunds cannot be mixed together

(a) Parallel structure with infinitives

Example – John likes to dance, to sing and to play the guitar.

(b) Parallel structure with gerund.

Example –

☑ **Right** : She likes reading, writing and composings poems

☒ **Wrong** : We should always find time for doing out chores, enjoying leisure time and to think of others.
☑ **Right** : We should always find time for doing our chores, enjoying leisure time and thinking of others.

☒ **Wrong** : Giving an oral presentation can required more preparation than to write a term paper.
☑ **Right** : Giving an oral presentation can require more preparation than writing a term paper.

Parallel structure with comparisons

Example –

1. I prefer reading to writing.
2. The climate of Agra is hotter than that of Shimla.

☒ **Wrong** : My book received a low grade more for its grammatical mistakes than what was said in it.
☑ **Right** : My books received a low grade more for its grammatical mistakes than for its contents.

Parallel structure with conjunctions

Such as

- Non only → ***but also***
- Either → ***or***
- Neither → ***nor***
- Both → ***and***

☒ **Wrong** : The workers disputed ***not only*** the magazine article ***but also*** they disputed the company's official statement.
☑ **Right** : The workers disputed ***not only*** the magazine article ***but also*** the company's official statement.

☒ **Wrong** : Sandra is neither from America nor Japan.
☑ **Right** : Sandra is neither from America nor ***from*** Japan.

Now read the following examples carefully

1. ☒ **Wrong** : The waiter suggested lamb chops, green beans and that we try to sweet potatoes.
   ☑ **Right** : The waiter suggested lamb chops, green beans, and sweet potatoes.
2. ☒ **Wrong** : She travelled to Agra, to Meerut and Bulandshahar.
   ☑ **Right** : She travelled to Agra, to Meerut, and ***to Bulandshahar***.
3. ☒ **Wrong** : A good speech must be audible, logical, and have clarity.
   ☑ **Right** : A good speech must be audible, logical and ***clear***.
4. ☒ **Wrong** : Surender likes hiking for great distances and to climb mountains.
   ☑ **Right** : Surender likes hiking for great distances and ***climbing mountains***.
5. ☒ **Wrong** : My wife likes imitating people and to tell jokes.
   ☑ **Right** : My wife likes imitating people and ***telling jokes***.

PRACTICE EXERCISE Parallelism

1. Eating a balanced diet (a) is more healthful (b) than junk food. (c) No error (d)

2. My desire is to visit (a) the museum and seeing (b) everything in the displays (c) No error (d)

3. He agreed to participate (a) in the talent show (b) but refused singing without a microphone. (c) No error (d)

4. The best beer has both (a) natural ingredients (b) and natural fermentation. (c) No error (d)

5. We should advertise the car wash (a) rather than (b) to be overlooked by potential customers. (c) No error (d)

6. I will either leave (a) from I.G. Airport (b) or from Dulles. (c) No error (d)

7. My history teacher is guilty of telling (a) about the past events (b) instead of an explanation of them. (c) No error (d)

8. The CEO has decided (a) to reduce costs, staff strength (b) and increase revenues. (c) No error (d)

9. The selling of books (a) is better (b) than selling tobacco. (c) No error (d)

10. Meera stopped at the grocery store (a) the bank (b) and ran to the library. (c) No error (d)

11. Nidhi likes to run (a) hunt, hike (b) and play. (c) No error (d)

12. Every night before bed time (a) Ansh Jainth reads a book, sings a song (b) and says a prayer. (c) No error (d)

13. He prefers (a) leasing or to buy (b) the equipment. (c) No error (d)

14. I know that (a) Monty will find working (b) with Sachin an enjoyable and worthwhile experience. (c) No error (d)

15. The yolk of one egg (a) contains 213 milligrams of cholesterol (b) and 5 grams of fat. (c) No error (d)

16. Neha likes (a) both who are popular (b) and who are not. (c) No error (d)

17. They have found a correlation (a) between exercise (b) earning good grades. (c) No error (d)

18. A good night's sleep (a) not only gives our body (b) a chance to rest but also energises us. (c) No error (d)

19. Cindy is happy (a) intelligent (b) and beautiful. (c) No error (d)

20. To run for an important political office (a) to manage a large organization (b) and practicing law effectively all require organizational and problem solving skills. (c) No error (d)

Answer Key ➤ Parallelism

1. (c)	2. (b)	3. (c)	4. (d)	5. (c)	6. (a)
7. (c)	8. (b)	9. (c)	10. (c)	11. (d)	12. (d)
13. (b)	14. (d)	15. (d)	16. (c)	17. (b)	18. (d)
19. (d)	20. (c)				

Explanation ➤ Parallelism

1. **(c)** Replace *than junk* food by than having junk food.
2. **(b)** Replace *and seeing* by and to see.
3. **(c)** Replace *refused singing* by refused to sing.
4. **(d)** No error
5. **(c)** Replace *to be overlooked* by be overlooked.
6. **(a)** Replace *I will either leave* by I will leave either.
7. **(c)** Replace *an explanation of them* by explaining them.
8. **(b)** Strengthen *staff should* be used at the place of staff strength.
9. **(c)** Replace *selling tobacco* by the selling of tobacco.
10. **(d)** Replace *and ran to the library* by and the library.
11. **(d)** No error
12. **(d)** No error
13. **(b)** Replace *or to buy* by buying.
14. **(d)** No error
15. **(d)** No error
16. **(c)** Replace *and who are...* by and those who are ...
17. **(b)** Replace *between exercise* by between exercising.
18. **(d)** No error
19. **(d)** No error
20. **(c)** Replace *and practicing law* by and to practice law.

CHAPTER 13 Conjunctions

Conjunctions

A conjunction is a word used to join two or more words / sentences or clauses.

Types of Conjunctions

1. Coordinating Conjunctions
2. Subordinating Conjunctions

Coordinating Conjunctions

Coordinating conjunctions are used to join two or more clauses of equal ranks. There are ***four types*** of coordinating conjunctions.

Cumulative or Copulative Conjunctions

and, as well as, both - and, not only - but also

Examples –

☒ **Wrong** : I love reading, writing, speaking, composing poems ***and to*** help others in my spare time.
☑ **Right** : I love reading, writing, speaking, composing poems, ***and*** helping others in my spare time.

☒ **Wrong** : You ***as well as*** I am listening to the music.
☑ **Right** : You ***as well as*** I are listening to the music.

☒ **Wrong** : Helen was both clever ***as well as*** beautiful.
☑ **Right** : Helen was both clever and beautiful.
☑ **Right** : Shakespeare was not only a poet but also playwright.

Adversative Conjunctions

but, still, yet
Adversative conjunctions are used to express contrast between two statements.

Examples –

1. He was ill, but he went to office.
2. He worked hard, yet he could not get pass marks.

☒ **Wrong** : She is very intelligent, and she cannot solve this question.
☑ **Right** : She is very intelligent, but she can not solve this question.

Illative Conjunctions

These conjunctions are used to show results.

So , therefore , for

Examples –

1. She was busy, so she could not attend the meeting.
2. I prefer to go by bus, for the train takes very long time.

Alternative Conjunctions

These conjunctions are used to express a choice between two or more alternatives.

or, else, nor, etc.

Examples –

1. Run fast, or you may miss the train.
2. Come in, or go out.
3. Neither Ram nor his friends live in this town.

Subordinating Conjunctions

A subordinating conjunction connects clauses of unequal grammatical rank. This conjunction introduces a clause that is dependent on the independent clause.

Examples –

1. Sandra said that her mother was smoking in the park.
2. Follow this road until you reach the station.

Types of subordinating conjunctions

1. Subordinating conjunctions of time:
   (when, while, before, after, till, since)
2. Subordinating conjunctions of place:
   (where, whereever)
3. Subordinating conjunctions of reason:
   (because, since, as)
4. Subordinating conjunctions of condition:
   (if, unless)
5. Subordinating conjunctions of concession:
   (though, although)
6. Subordinating conjunctions of result or consequence:
   (so, that, such, that)
7. Subordinating conjunctions of purpose:
   (in order that, lest, so that)
8. Subordinating conjunctions of comparison:
   (than, as ..... as, so ..... as)
9. Subordinating conjunctions of manner:
   (as if, as though)

Rules Regarding Conjunctions

1. Coordinating conjunctions are applied to join same parts of speech.
2. Some conjunctions go with their fixed pair, learn them such as
   - No sooner .... than
   - Hardly / scarcely .... when / before
   - As .... as
   - So .... as
   - Whether .... or
   - Lest .... should
   - Though / although .... yet / ,
   - Between ....and

   In modern English, we usually use comma with though / although.
3. When **WH** words ***what***, ***when***, ***where***, ***why*** etc are used as a conjunction, they are followed by a subject.

   Examples –

   ☒ **Wrong** : I know the reason ***why was*** she making a noise.

   ☑ **Right** : I know the reason ***why she*** was making a noise.
4. ***'Asked'*** is not followed by conjunction ***'that'*** we can use ***if/whether*** or ***WH*** word.

   Examples –

   ☒ **Wrong** : He ***asked*** me that if I know the way to station.

   ☑ **Right** : He ***asked*** me if I knew the way to station.
5. With ***'doubt'*** in positive sentence apply ***if/wether***, and in negative sentence apply ***'that'***.

   Examples –

   ☒ **Wrong** : I doubt that he can help me.

   ☑ **Right** : I doubt if he can help me.

   ☑ **Right** : I doubt whether he can help me.

   ☑ **Right** : I do not doubt that she can solve this question.

Errors in the use of Conjunctions

Rule : 1 ***As well as***

In case of ***as well as*** the verb is applied according to the first subject.

☒ **Wrong** : Candida ***as well as*** I have composed an elegy.

☑ **Right** : Candida ***as well as*** I has composed an elegy.

NOTE ***'As well as'*** can link items other than nouns and pronouns, as long as they match each other grammatically.

Examples –

1. He directs plays as well as writes them.
   (directs – verb; writes – verb)
2. I was surprised as well as pleased to see her melting in his arm.
   (surprised – adjective; pleased – adjective)

Rule : 2 *Lest .... should*

'Lest' means it may not happen. *'Lest'* is a negative conjunctions, so it does not require any negative word.

☒ **Wrong** : Walk carefully lest you may fall.
☑ **Right** : Walk carefully lest you should fall.

☒ **Wrong** : Work hard lest you should not fail.
☑ **Right** : Work hard lest you should fail.
OR Work hard lest you fail.

Note : Should can be omitted.

Rule : 3 *No sooner .... than*

1. No sooner is a comparative degree, so it is followed by than,
2. Do not use then
3. Do not use any negative word with the statement that contains no sooner.
4. We use inversion with no sooner ..... than.

Inversion means putting helping verb before the subject.

[No sooner + did + subject + V_1 + object + than + subject + V_2 + object]

OR

[No sooner + had + subject + V_3 + object + than + subject + V_2 + object]

☒ **Wrong** : No sooner she had entered the office than the boss began to shout at her.
☑ **Right** : No sooner had she entered the office than the boss began to shout at her.

☒ **Wrong** : No sooner did he reach the station when it began to rain.
☑ **Right** : No sooner did he reach the station than it began to rain.

Rule : 4 *Hardly / Scarcely .... when / before*

[Hardly / scarcely + had + subject + V_3 + object + when + subject+ V_2 + object]

OR

[Hardly / scarcely + did + subject + V_1 + object + when + subject+ V_2 + object]

☒ **Wrong** : Hardly had the teacher entered the classroom than all the students stood up.
☑ **Right** : Hardly had the teacher entered the classroom when all the students stood up.

Rule : 5 *Although / though .... yet /,*

Do not use 'but' with it.

☒ **Wrong** : Although she worked hard but she could not get through knowledge of mathematics.
☑ **Right** : Although she worked hard yet she could not get through knowledge of mathematics.

Rule : 6 *Where / In which*

Use *'where'* to refer to a specific location, and for other cases use *'in which'*.

☒ **Wrong** : The town in which I was born is known for its historical building.
☑ **Right** : The town where I was born is known for its historical building.

☒ **Wrong** : The company where I work has gone bankrupt.
☑ **Right** : The company in which I work has gone bankrupt.

Rule : 7 *Whether .... or / if*

In case of introducing alternative possibilities, usually with ***'or not'*** we use ***whether (not if)***

☒ **Wrong** : I cannot decide if to study or to go out with my wife.
☑ **Right** : I cannot decide whether to study or to go out with my wife.

We can use ***if*** at the place of ***whether***, but there are certain points in which you cannot use ***'if'***

Whether is obligatory in the following cases:

After a preposition —

☒ **Wrong** : They quarrelled about if he should have the operation or not.
☑ **Right** : They quarrelled about whether he should have the operation or not.

Before infinitives —

☒ **Wrong** : We didn't know if to laugh or cry.
☑ **Right** : We didn't know whether to laugh or cry.

Rule : 8 ***Because / Because of*** and due to

1. ***'Because'*** is a conjunction which is used to introduce an ***'adverb clause'***.
2. If a sentence begins with ***because***, we cannot use ***so / therefore*** in it.

☒ **Wrong** : Because the weather was cold, so we stayed home.
☑ **Right** : Because the weather was cold, we stayed home.

Rule : 9 The word reason is not followed by ***'because'***

☒ **Wrong** : The reason why she did not go to office was because her mother was ill.
☑ **Right** : The reason why she did not go to office was that her mother was ill.

Rule : 10 ***Unless, till, untill,*** are not followed by ***'not'***

☒ **Wrong** : Unless he does not come to me, I will not help him.
☑ **Right** : Unless he comes to me, I will not help him.

Rule : 11 ***Like / as***

'Like' is used as a preposition, while ***'as'*** is a conjunction.

'Like' is used to compare only noun, (i.e. use ***'like'*** to say two things are similar)

Examples —

The earth like other planets spins on an axis.

'As' is used to compare clauses. (i.e. use ***'as'*** if two actions are similar)

Examples —

A globe spins around an axis as does the earth itself.

PRACTICE EXERCISE Conjunctions

Instructions : Out of the following four sentences, select the most suitable sentence with respect to grammar and usage.

1. (a) I know why was he making a noise.
(b) I know why he was making noise.
(c) I know why he was making a noise.
(d) I know why were he making a noise.

2. (a) I doubt that he will come tomorrow.
(b) I doubt so he will come tomorrow.
(c) I doubt if will he come tomorrow.
(d) I doubt if he will come tomorrow.

3. (a) She is very intelligent, so she cannot solve these sums.
(b) She is very intelligent, therefore she cannot solve these sums.
(c) She is very intelligent, but she cannot solve these sums.
(d) She is very intelligent that she cannot solve these sums.

4. (a) He was ill, but he went to office.
(b) He was ill, so he went to office.
(c) He was ill so that he went to office.
(d) He was ill because he was going to office.

5. (a) Hardly he had reached the station than it begain to rain.
(b) Hardly he reached the station than it began to rain.
(c) Hardly did he reached the station than it began to rain.
(d) Hardly had he reached the station when it began to rain.

6. (a) I asked him if he knows the way to post office.
(b) I asked him that he knew the way to post office.
(c) I asked him if he knew the way to post office.
(d) I asked him that if he knew the way to post office.

7. (a) No sooner did she enter the room when all the people stood up.
(b) No sooner had she enter the room than all the people stood up.
(c) No sooner had she entered the room than all the people stood up.
(d) No sooner did she entered the room then all the people stood up.

8. (a) Although he worked hard, but he could not pass.
(b) Although he worked hard, but he cannot pass.
(c) Although he worked hard, yet he could not pass.
(d) Although he worked hard, so he could not pass.

9. (a) Shakespeare was both a poet and a playwright.
(b) Shakespeare was both a poet and playwright
(c) Shakespeare was both a poet as well as playwright.
(d) Shakespeare was both a poet along with playwright.

10. (a) Neither Ram nor his friends is coming.
(b) Neither Ram nor his friends are coming.
(c) Neither Ram or his friends are coming.
(d) Neither Ram nor his friends seems to be coming.

Answer Key ➤ Conjunctions

1. (c)	2. (d)	3. (c)	4. (a)	5. (d)	6. (c)
7. (c)	8. (c)	9. (b)	10. (b)		

Explanations > Conjunctions

1. **(c)**

Option (a) is wrong because why is followed by helping verb.

Option (b) is wrong because the correct phrase is make a noise.

Option (c) is correct because why is followed by subject which is a correct sequence.

Option (d) is wrong because why is followed by were, which makes no sense at all.

2. **(d)**

'Doubt' is followed by if or whether so options (a) and (b) are wrong.

Option (c) is wrong because it is followed by helping verb.

Option (d) is the best answer choice.

4. **(a)**

Options (b), (c) and (d) have been connected with the wrong conjunctions, so they are wrong.

Option (a) makes proper sense, so option (a) is the best answer choice.

5. **(d)**

If a sentence being with a negative adverb inversion is used, so options (a) and (b) are wrong.

Hardly is followed by when, so option (c) is wrong.

Rule-1

[Hardly + had + subject + V_3 + object + when + subject + V_2 + object]

According to rule discussed above (d) is the best answer choice.

6. **(c)**

If the principal clause is in the past tense, the sub-ordinate clause should be in the past tense, so option (a) is wrong.

Asked is not followed by conjunction that so option (b) is wrong.

Option (d) is wrong because if and that have been used together.

Option (c) is the best option choice.

7. **(c)**

No sooner is followed by than, so option (a) is wrong.

Had takes past participle (V_3) so option (b) is wrong.

Did takes V_3 form of the verb, so option (d) is wrong.

Rule-1

[No sooner + had + subject + V_3 + object + than + subject + V_3 + object]

Option (c) is the best option choice.

8. **(c)**

Although is not followed by but, so options (a) and (b) are wrong.

Although is not followed by so, so option (d) is wrong.

Although is followed by yet, so (c) is the best option choice.

9. **(b)**

Option (a) is wrong because article a has been used twice.

Both is followed by and, so options (c) and (d) are wrong.

Option (b) is the best option choice.

10. **(b)**

In case of neither - nor, the verb is used according to the second subject, so option (a) has verb error.

Neither is followed by nor, so option (c) wrong.

Option (d) also has verb error.

Option (b) is the best option choice.

■

CHAPTER 14

Prepositions

Prepositions

Prepositions are applied to show the relationship between a noun or a pronoun and other element in a sentence or a phrase.

EXAMPLES —

1. Belinda is sitting at the desk.
2. Transfer this cheque in my name.
3. Sandra has good command of phonetics.
4. She gave me a cheque for Rs. 20,000.

Classification of Prepositions

1. Prepositions that indicate location

above, behind, below, beside, between, under, over, infront of, etc.

EXAMPLES —

1. He is at the table.
2. The monkey was sitting in the tree.

2. Prepositions that indicate movement or direction:

These prepositions tell us the movement of someone or something.
along, at , by, from , in, into, off, to, towards or toward etc.

EXAMPLES —

1. They walked along the river.
2. She was coming from Kanpur.
3. I saw them going towards the temple.

3. Prepositions that indicate time

after, at, before, during, for, since, on, in, from, it until, etc.

EXAMPLES —

1. You can meet me at 6 o'clock.
2. I have been reading this book since morning.
3. You can meet me on Monday.

4. Compound Prepositions

because of, by way of, in front of, with respect to, upto, in spite of, in reference to, with a view to, on account of, according to etc.

EXAMPLES —

1. He comes here with a view to learning English.
2. There is a tree in front of my house.
3. According to Mohan, she is a good student.

Usage of Prepositions

1. In / Into

In 

Into  ***motion***

1. In indicates state position.
2. Into indicates motion.

EXAMPLES —

1. My books are in my bag.
2. The cat fell into the well.

☒ **Wrong** : The teacher asked the students to keep their books in their bags.
☑ **Right** : The teacher asked the students to keep their books into their bags.

2. In / At : (time)

In the morning In the evening In the afternoon at noon at night at midnight	At down At dusk, at sunrise At sunrise

At indicates fixed time

- You can met me at 9:30.

In : We use ***'in'*** before a longer period of time, such as ***month, year, century***.

EXAMPLES —

1. In March
2. In 2019
3. In the 19th century

In / At : (Place)

'In' is used before big cities.
'At' is used before small cities.'
'At' is also used for specific addresses.

EXAMPLES —
1. Christy lives at Laxmi Nagar a small town in Delhi.
2. You can send this letter *'at'* B-21120, Street no. 5, Harsh Vihar, Delhi-110093.

Special usages of (IN)

☒ **Wrong** : Please write with pencil.
☑ **Right** : Please write in pencil.

in +	Pencil Ink Pen	with +	a Pencil a Pen

☑ **Right** : Write with a pen.

IN + Name

☒ **Wrong** : Transfer this cheque on my name.
☑ **Right** : Transfer this cheque in may name.
☑ **Right** : There is no property in my name.
☑ **Right** : In the name of god, I pronounce you man and wife.

3. Usage of ON

1. ***day :*** on Sunday, on Monday, etc.
2. ***date :*** on 15 August, on 16 March, etc.
3. ***festival :*** On Holi, on Diwali
1. ***On :*** is also used to indicate position. ***(Physically in contact with something)***

   EXAMPLES —
   The ball is ***on*** the table.

2. ***On*** is also used when we talk about
   1. Subject matter 2. Horizontal or vertical surface

   EXAMPLES —
   1. I am going to deliver a lecture on morphology.
   2. Keep it on the table.
   3. Can you paste this photo on the wall.

EXAMPLES —
1. I was sitting ***under*** the tree.
2. The monkey was sitting ***in*** the tree.
3. The aeroplane was ***over/above*** the tree.
4. There is a tree ***infront*** of my house.

4. Of / off

'Of' means belonging to something.

EXAMPLES —
1. The pages ***of*** this book seem to be torn by him.
2. Can you give me a glass ***of*** water?

'Off' means away from something?

EXAMPLES —

1. He took him feet off the desk.
2. Don't try to get on or off a moving train.
3. Take your hands off my cheeks.

5. Below and under

Below and under often have the same meaning 'lower than'.

EXAMPLES —

1. The restaurant kitchen was ***below*** the living room.

   or

   The restaurant kitchen was ***under*** the living room.

   But we use ***'below'*** and not ***'under'*** when we mean lower than, but not necessarily directly under.
2. ☒ **Wrong** : I could see the park ***under*** us.

   ☑ **Right** : I could see the park ***below*** us.

> We also use ***'under'*** when we mean ***'covered by'***
> - The dog was sleeping ***under*** the table.

6. To, Towards, From

Both ***'To'*** and ***'Towards or Toward'*** are the prepositions of direction.

'To' is used to indicate destination.

EXAMPLE —

1. He went ***to*** school yesterday.

'Towards' is used to indicate direction.

EXAMPLE —

1. I saw him going towards the temple
   OR
   I saw him going toward the temple.

'From' is used with the point of departure.

EXAMPLE —

1. I came to meet you from Delhi.
2. She has been transferred from this office.

'From' is also applied to indicate the starting point of an action.

EXAMPLE —

1. You can come to meet me from 10 am to 11 am.

7. Across

'Across' means from one side to the other of something with clear limits. Such as an area of land, a road, or a river.

EXAMPLE —

1. I saw Candida walking ***across*** the river.
2. The stream was small enough to jump ***across***.
3. My grandparents used to live ***across*** the street.

8. Beside / Besides

'Beside' means by the side of or, next to.

EXAMPLES —

1. I saw a house beside a small lake.
2. She came and sat beside me.

Besides means in addition to

EXAMPLES —

1. I want to learn other languages besides French and Spanish.
2. Besides its famous cakes, the bakery also makes delicious cookies.

'With' is used before instrument, such knife, gun, stick etc.

- A lion was killed by Aman ***with*** a gun.

'Since' Indicate point of time.

EXAMPLE —

- She has been reading this book ***since*** morning.

'For' Indicates period of time.

EXAMPLES —

- We have been watering the plants ***for*** two hours.
  For' also means to exchange.
- I bought this book for Rupees 500.
  For is also used to indicate a purpose or reason.

EXAMPLES —

1. Would you like a bag ***for*** those books?
2. What is she smiling ***for***?

9. Among/Between

'Among' means in the middle of or surrounded by other things. ***Among*** is used for more than two people or things.

EXAMPLES —

1. I saw a few familiar faces ***among*** the crowd.
2. These disease are more common ***among*** young children?

'Between' means in or into the space that separates two places, people or things.

Between is most often used to introduce a prepositional phrase which contains two singular or plural noun phrases. ***Between*** is followed by and.

EXAMPLES —

1. Distribute these apples ***between*** Ram and Mohan.
2. You can come to meet me ***between*** 6 and 7.

We can also use ***'between'*** for more than two objects, but they should belong to the same category. We also use ***between*** when we list each item individually.

- The match will be played ***between*** India, England and Pakistan.

10. Despite/Inspite of:

Despite and ***Inspite of*** mean the same i.e. on the contrary.

'Despite' is never followed by ***'of'***.

EXAMPLE —

☒ **Wrong** : Despite of working hard, she could not get good marks.

☑ **Right** : Despite working hard, she could not get good marks.

or

Inspite of working hard, she could not get good marks.

11. Over

'Over' means 1. At a higher level.

2. More than certain amount or measurement.

EXAMPLES —

1. The tree formed a canopy ***over*** my head.
2. You are supposed to pay me ***over*** Rs. 5000 an hour.

12. According to

'According to' means in the opinion of others. We do not use me/us with it.

☒ **Wrong** : According to me she is a good student.

☑ **Right** : In my opinion she is a good student.

13. Behind

Behind: At the back (of) It is used to show the position of an object placed in the back side of a particular place.

EXAMPLES —

1. My friend was driving the car while I was sitting ***behind***.
2. The temple is ***behind*** a big banyan tree.
3. I hung my coat ***behind*** the door.

14. During

'During' means all through a period of time. ***'During'*** is used to say when something happens, and for answering the quotation 'how long?'

EXAMPLES —

1. My elder son was born ***during*** a thunderstorm.
2. I have to have my window open ***during*** the night (means the whole of night).
3. No flash photography is permitted ***during*** the concert.

15. Throughout

'Throughout' means the whole period of time.

EXAMPLE —

1. I thank you for all the support ***throughout*** the years.

16. Except and Except for

'Except' means excluding. We use ***'except'*** to rule out an item within the group that our statement applies to.

EXAMPLES —

1. Everyone ***except*** me thought that she was a good student.
2. They are interested in nothing ***'except'*** watching television.

'Except for' means also excluding.

EXAMPLES —

1. ***Except*** for Tina, everyone was pleased.
2. ***Except*** for Lousia, who is away in London this weekend, we will all be at the party.

NOTE : We can use ***except*** and ***except for*** interchangeably, however ***'except for'*** not ***'except'*** must be used when it is the part of a phrase beginning a sentence.

☑ **Right** : I like all fruit except oranges. ***or*** I like all fruit except for oranges.

☒ **Wrong** : Except me, everyone thought, it, was a good idea.

☑ **Right** : Except for me, everyone thought, it was a good idea.

17. In between

'In between' means in a place that is between two or other things or places.

EXAMPLES —

1. In between leaving her job and doing a degree, she managed to have another child.
2. We stopped for lunch somewhere in between Anupshahar and Bulandshahar.

18. For the sake of

'For the sake of' means inorder to achieve something.

EXAMPLE —

1. For the sake of your health, you should give up smoking.

19. 'But for'

'But for' means without. ***But for***, is used to talk about situations that would have happened if something had not existed or had not been true.

1. But for your knowledge of first aid, I would not have survived.

Rules Regarding Prepositions

Rule : 1 A preposition needs an ***objective noun or pronoun***.

EXAMPLES —

1. He will go with ***them***.
2. ***To*** whom should I give this book?
3. When you rang me up, I was talking ***to*** Mohan.

☒ **Wrong** : This problem is so complicated that it cannot be solved by we people.

☑ **Right** : This problem is so complicated that it cannot be solved by us people.

Rule : 2 Always use ***gerund*** after prepositions.

EXAMPLES —

1. I am fond ***of composing*** poems.
2. Sandra is smart ***at cooking*** food.

☒ **Wrong** : I am looking forward to meet you.
☑ **Right** : I am looking forward to meeting you.

☒ **Wrong** : She is used to drink wine with them.
☑ **Right** : She is used to drinking wine with them.

Rule : 3 If there is a preposition in ***active voice***, the same preposition should be in ***passive voice*** too.

EXAMPLES —

Active voice : He looked after his parents.
Passive voice : His parents were looked after by him.

Active voice : You must listen to my words
Passive voice : ☒ **Wrong** : My words must be listened by you.
☑ **Right** : My words must be listened to by you.

Rule : 4 Always use the preposition of movement ***'to'*** with motion verbs such as ***go, walk, run, move*** etc.

EXAMPLES —

1. He want to America ***to*** meet one of this girlfriends.
2. Drive ***to*** the country if you want to see a bevy of beautiful women.

Do not use ***to before*** home and abroad.

☒ **Wrong** : Candida went to home to meet her mother.
☑ **Right** : Candida went home to meet her mother.

'Arrive' is not a motion verb because it indicates no motion. Use the preposition 'at' after the verb arrive.

- The train arrived at the station ten minutes ago.

Rule : 5 There are several ***verbs*** that may have ***more than one preposition***.

1. Shout to/Shout at

Shout to is used, when you raise your voice to attract someone's attention.
Shout at is used, when you raise your voice because you are angry with someone.

EXAMPLES —

Shout to

2. She shouted to me from the upstairs window.
3. I shouted to them across the busy road, but they didn't hear me.

2. Shout at

EXAMPLES —

1. The receptionist decided to leave because the boss often shouted at her.
2. Please do not shout at her.

3. Command of/Command ever

'Command of' is used before a subject. ***'Command over'*** is used before people.

EXAMPLES —

☒ **Wrong** : He has an excellent ***command over*** mathematics.
☑ **Right** : He has an excellent ***command of*** mathematics.
☒ **Wrong** : He has ***command of*** his children.
☑ **Right** : He has ***command over*** his children.

4. **Born of/Born in**

Born of + Parents / *Born + in* + family, village, country etc.

EXAMPLES —

☒ Wrong : He was *born in* poor parents.
☑ Right : He was *born of* poor parents.
☑ Right : Dr. Faustus was *born in* Germany.

5. **Angry with/Angry at**

Angry with + a person / *Angry at* + something

EXAMPLES —

1. He was angry with me.
2. Sheela was angry at my rude behaviour.

6. **Abide by/Abide with/Abide in**

Abide by – To obey the law.

EXAMPLE —

- We should abide by the rules of traffic.

Abide with – To remain faithful.

EXAMPLE —

- We should abide with our friends.

Abide in - To stay

EXAMPLE —

- We should abide with our friends.

7. **Made of/Made from/Made out of/Made with**

Made of used when we talk about the basic material of something. The meaning of 'made of' is composed of.

EXAMPLE —

- I saw Lydia wearing a beautiful necklace made of gold.

Made from is usually used when we mean chemical change.

EXAMPLE —

- Curd is *made from* milk.

Made out of - We usually use *made out* of when we talk about something that has been changed or transformed from one thing into another.

- They were living in tents made out of old plastic sheets.

Made with - 'Made with' is used when we talk about the ingredients of food and drink.

- This dish is made with mutton, red peppers and herbs.

8. **Blind to/Blind In**

'Blind to' means to ignore someone.

EXAMPLE —

- You should not blind to that stunning woman.

'Blind in' + eye

EXAMPLE —

- He was *blind in* one eye.

9. **Cheque of/Cheque for**

Cheque of is a wrong use of preposition. ***say Cheque for***

☒ **Wrong** : Sandra gave me a cheque of Rs. 50,000.

☑ **Right** : Sandra gave me a cheque for Rs. 50,000.

10. **Differ with/Differ from**

Rule : 6 Always use ***required preposition***, when an infinitive is used to ***indicate purpose***.

☒ **Wrong** : He has a beautiful house to live.

☑ **Right** : He has a beautiful house to live ***in***.

☒ **Wrong** : John gave me an elegant shirt to put.

☑ **Right** : John gave me an elegant shirt to put ***on***.

☒ **Wrong** : Please give him a handed chair to sit.

☑ **Right** : Please give him a handed chair to sit ***in***.

Rule : 7 The following verbs are not following by any prepositions in Active Voice.

Enter, Attack, Discuss, Explain, Depict, Enunciate, Tell, Request, Order, Marry, Divorce, Emphasise, Sign, Match, Resemble, Answer, Phone, Meet, Contact, Call, Question, Stress, Comprise, Investigate, Resist, Affect.

1. ☒ **Wrong** : Wearing a mild smile on her rosy lips Cathy entered ***into*** my room.
   ☑ **Right** : Wearing a mild smile on her rosy lips, Cathy entered my room.
2. ☒ **Wrong** : You should emphasize on the rules of grammar while writing something.
   ☑ **Right** : You should emphasize the rules of grammar while writing something.
3. ☒ **Wrong** : I will answer to your question when I finish eating.
   ☑ **Right** : I will answer your question when I finish eating.
4. ☒ **Wrong** : Rochester married to Jane.
   ☑ **Right** : Rochester married Jane.
   But we can say
   Jane was married to Rochester.
5. ☒ **Wrong** : I hope to form a team that comprises of six members.
   ☑ **Right** : hope to form a team that comprises six members.

NOTE

Explain, complain talk, listen, write and ***speak*** are followed by ***preposition 'to'*** if they have *objects*.

- I will speak to you later.

Rule : 8 If a sentence contains two or more words that require prepositions.

Use appropriate prepositions with each of the words if they require different preposition with them.

☒ **Wrong** : She is senior and more experienced from Meera.

☑ **Right** : She is senior ***to*** and more experienced than Meera.

☒ **Wrong** : Amphibians can live in water as well as land.

☑ **Right** : Amphibians can live *in* water as well as *on* land.

☑ **Right** : She is annoyed *with* and indifferent *to* Mohan.

Rule : 9 When two or more words in a sentence require the same preposition. Use the preposition only once.

Wordy : He was desirous of and confident of success.

Better : He was desirous and confident of success.

Rule : 10 Repeat the preposition when there are several parallel phrases connected by than or as well as.

☒ **Wrong** : My wife thought more of her home than my parents.

☑ **Right** : My wife thought more of her home than of my parents.

Verbs with following Prepositions	
(On)	
1.	Comment on
2.	Concentration on
3.	Count on (= rely)
4.	Depend on
5.	Embark on
6.	Live on (food)
7.	Operate on (someone)
(At)	
8.	Arrive at
9.	Guess at
10	Hint at
11.	Marvel at
12.	Play at (Pretend)
13.	Wonder at
14.	Work at
15.	Look at
16.	Peep at
17.	Peer at

Verbs with following Prepositions	
(Against)	
18.	Prejudiced against
19.	Insure against
20.	Protest against (or to)
21.	React against (or to)
22.	Warn against (or about)
(About)	
23.	Joke about
24.	Laugh about
25.	Think about
26.	Worry about
27.	Tell about something
(In)	
28.	Believe in
29.	Delight in
30.	Engaged in
31.	Included in
32.	Take a pride in
33.	Invest in
34.	Persist in
35.	Involved in

Verbs with following Prepositions	
(From)	
36.	Abstain from
37.	Refrain from
38.	Borrow from
39.	Demand something from
40.	Differ from
41.	Draw money from
42.	Hinder from
43.	Prevent from
44.	Suffer from
45.	Recover from
46.	Prohibit from
47.	Emerge from
48.	Exclude from
50.	Distinguish A from B.
(For)	
51.	Apologize for
52.	Blame someone for
53.	Beg for
54.	Hope for
55.	Pay for
56.	Prepare for
57.	Provide for
58.	Search for
59.	Vote for
60.	Wait for
(Of)	
61.	Accuse of
62.	Approve of
63.	Beware of
64.	Consist of
65.	Convicted of (a crime)
66.	Dream of

Verbs with following Prepositions	
67.	Repent of
68.	Smell of
69.	Taste of
70.	Suspect of
71.	Think of
72.	Get third of
73	Warn of
(To)	
74.	Accustomed to
75.	Amount to
76.	Appeal to
77.	Apply to
78.	Attached to
79.	Belong to
80.	Confined to
81.	Consent to
82.	Converted to
83.	Listen to
84.	Object to
85.	React to
86.	Reduced to
87.	Resort to
88.	Respond to
89.	Reply to
90.	Subjected to
91.	Key to
92.	Submit to
93.	Succumb to
94.	Surrender to

Verbs with following Prepositions	
95.	Subscribe to
96.	Talk to
97.	Turn to
98.	Yield to
99.	Prone to
100.	Want to
(With)	
101.	Agree with someone
102.	Charged with (crime)
103.	Compete with
104.	Comply with

Verbs with following Prepositions	
105.	Conflict with
106.	Cope with
107.	Infested with
108.	Part with
109.	Quarrel with
110	Fight with
111.	Satisfied with
112.	Reason with someone
113.	Combat with difficulties
114.	Begin with
115.	Conform with one's views

PRACTICE EXERCISE Prepositions

Instructions : Out of the following four sentences, select the most suitable sentence with respect to grammar and usage.

1. (a) Ansh has good command on mathematics.
(b) Ansh has good command over mathematics.
(c) Ansh has good command upon mathematics.
(d) Ansh has good command of mathematics.

2. (a) Your trousers do not match with your shoes.
(b) Your trousers do not match at your shoes.
(c) Your trousers do not match your shoes.
(d) Your trousers do not match over your shoes.

3. (a) This book comprises over three hundred pages.
(b) This book comprises of three hundred pages.
(c) This book comprise three hundred pages.
(d) This book comprises three hundred pages.

4. (a) You can come to meet me between 9 to 10.
(b) You can come to meet me between 9 and 10.
(c) You can come to meet me between from 9 to 10.
(d) You can come to meet me from 9 and 10.

5. (a) The monkey was siting over a tree.
(b) The monkey was sitting on a tree.
(c) The monkey was siting in a tree.
(d) The monkey was sitting at a tree.

6. (a) The teacher asked the student to keep his books in his bag.
(b) The teacher asked the student to keep his books into his bag.
(c) The teacher asked the student to keep his books with his bag.
(d) The teacher asked the student to keep his books on his bag.

7. (a) Please write with black ink.
(b) Please write in black ink.
(c) Please write at black ink.
(d) Please write into black ink.

8. (a) She bought this book with Rs. 200.
   (b) She bought this book in Rs. 200.
   (c) She bought this book by Rs. 200.
   (d) She bought this book for Rs. 200.

9. (a) Sheela was sitting over the desk.
   (b) Sheela was sitting upon the desk.
   (c) Sheela was sitting on the desk.
   (d) Sheela was sitting at the desk.

10. (a) I am looking forward meeting you.
   (b) I am looking forward to meeting you.
   (c) I am looking forward to meet you.
   (d) I am look forward to meet you.

Answer Key ➤ Prepositions

1. (d)	2. (c)	3. (d)	4. (b)	5. (c)	6. (b)
7. (b)	8. (d)	9. (d)	10. (b)		

Explanations ➤ Prepositions

1. (d)

Command on is a wrong phrase, so option (a) is wrong.

Command over is a wrong phrase, so option (b) is wrong.

Command upon makes no sense, so option (c) is wrong.

Thus, option (d) is the best option choice.

2. (c)

The verb 'Match' does not require any preposition, so options (a), (b) and (d) are wrong.

Thus, option (c) is the best option choice.

3. (d)

The verb 'comprise' does not require any preposition, so options (a) and (b) are wrong.

Option (c) is wrong because it has verb error.

Thus, option (d) is the best option choice.

4. (b)

Between is followed by and, so option (a) is wrong.

Between and from cannot be used together, so option (c) is wrong.

From with and make no sense, so option (d) is wrong.

Thus option (b) is the best option choice.

5. (c)

The correct prepositional phrase is ***sitting in a tree***, so options (a), (b) and (d) are wrong.

Thus, option (c) is the best option choice.

6. (b)

To indicate motion we use the preposition into, so options (a), (c) and (d) are wrong.

Thus option (b) is the best option choice.

7. (b)

The correct prepositional phrase is write in black ink, so options (a), (c) and (d) are wrong.

Thus (b) is the best option choice.

8. (d)

The correct prepositional phrase is book for Rs. 200.

Thus option (d) is the best option choice.

9. (d)

The correct prepositional phrase is at the desk, so options (a), (b) and (c) are wrong.

Thus, option (d) is the best option choice.

10. (b)

The Rule says :

Look forward to + V_1 + ing + object

According to the above mentioned rule, the correct prepositional phrase is ***looking forward to meeting you***, so options (a), (c) and (d) are proved to be wrong.

Thus, option (b) is the best option choice.

CHAPTER 15 Fill in the blanks

Set-1 Fill in the blanks

1. Sonia disliked eating in the ________ restaurant.
(a) nosily (b) noiseful
(c) noisy (d) noised

2. He was convicted ________ manslaughter.
(a) of (b) from
(c) for (d) to

3. How much did you spend ________ the roses
(a) in (b) on
(c) of (d) from

4. Reaction to Jane Austen's novels has oscillated between ________ and condescension.
(a) indifferent (b) disapproval
(c) dismissal (d) adoration

5. He does not ________ any fixed principle.
(a) go through (b) go about
(c) go with (d) go upon

6. As soon as he got the telegram, he ________ in a taxi.
(a) would have left (b) left
(c) had left (d) has left

7. We had difficulty ________ finding a parking place
(a) at (b) about
(c) in (d) for

8. The human mind seems to have built in ________ against original thought.
(a) interests (b) safeguards
(c) prejudices (d) ideas

9. I am used to ________ hockey with my friends.
(a) plays (b) playing
(c) played (d) have been playing

10. When I saw him he ________ watering the plants.
(a) is (b) are
(c) was (d) were

11. The deceased left ________ him two young children
(a) with (b) by
(c) behind (d) for

12. Many women in developing countries experience a cycle of poor health that ________ before they are born.
(a) originates (b) derives
(c) begins (d) establishes

13. I asked him if I ________ borrow his car for a day.
(a) will (b) could
(c) can (d) should

14. They have lost ________ of their contacts.
(a) little (b) much
(c) many (d) small

15. I praise ________ honesty of this boy.
(a) a (b) an
(c) the (d) your

16. I have to recite Tagore's sonnet. I will be the ________ one to do so in my school.
(a) earliest (b) initial
(c) solitary (d) only

17. Our rich culture inspires us to take ________ in our heritage.
(a) prided (b) proudly
(c) pride (d) proud

18. The car broke down and we ________ get a taxi
(a) were to (b) are to
(c) had to (d) have to

19. She is kind enough to ______
(a) could (b) can
(c) will (d) would

20. He was junior ________ me in college and we have been friends since then.
(a) with (b) than
(c) from (d) to

21. I have never seen ________ animal before.
(a) any larger (b) as large
(c) such a large (d) so large

22. He was sorry ________ late last night.
(a) about coming (b) to come
(c) for coming (d) from coming

23. Vinod, two of ________ brothers attend primary school, wishes to be a school teacher after two graduation from the university.
(a) whom (b) them
(c) whose (d) which

24. How did these things come ________.
(a) about (b) into
(c) at (d) down

25. The news was ________ good to be true.
(a) very (b) too
(c) so (d) as

Set-2 Fill in the blanks

1. Ramu is in the habit of revising just once, but Suresh never understands a subject ________ he revises it for atleast three times.
(a) In case (b) thus
(c) although (d) unless

2. He was sorry ________ later last night.
(a) about coming (b) to come
(c) at coming (d) for coming

3. The sparrows took no ________ the bread.
(a) notice of (b) notice about
(c) notice from (d) notice to

4. He shows great ability ________ Mathematics.
(a) at (b) for
(c) about (d) in

5. Poets have compared this world ________ a stage.
(a) on (b) off
(c) over (d) to

6. You should not confide ________ a stranger.
(a) in (b) on
(c) over (d) of

7. The telephone kept ________ ringing, but no one attended.
(a) off (b) on
(c) up (d) up with

8. I prevailed ________ him to contest the election.
(a) on (b) off
(c) over (d) of

9. A.R. Rahman ________ the music for commonwealth Games 2010.
(a) song (b) made
(c) composed (d) produced

10. All is not well ________ the automobile sector
(a) of (b) down
(c) in (d) to

11. Out teacher encouraged ________ setting the poem to music.
(a) his (b) him
(c) he (d) he's

12. The valley is known for its ________ growth of vegetation.
(a) luxurious (b) luxury
(c) luxuriant (d) luxuriously

13. My Murugan has been is this college ________ 2010.
(a) for (b) since
(c) after (d) before

14. Nobody ________ on time to attend the seminar.
(a) come (b) comes
(c) came (d) coming

15. It is about time we ________ to school.
(a) going (b) go
(c) goes (d) went

16. Jone is annoyed ________.
(a) on me (b) with me
(c) against me (d) over me

17. He slipped ________ his old ways and started drinking again.
(a) into (b) off
(c) by (d) in

18. Many premier educational institutions come forward to have a ________ with flourishing industries.
(a) tie down (b) tie up
(c) tie in (d) tie on

19. She is kind enough to
(a) helps (b) helped
(c) helping (d) help

20. How long had you ________ this car?
(a) have (b) has
(c) had (d) having

21. Her office is next ________ ours.
(a) to (b) on
(c) with (d) from

22. Would you like to purchase ________ edition of this play?
(a) revise (b) the revise
(c) the revised (d) revised

23. Would you mind ________ going out?
(a) me (b) him
(c) my (d) I

24. I saw him ________ a detective novel.
(a) reading (b) being read
(c) have been reading (d) was reading

25. The two sisters should not quarrel with ________ other.
(a) one (b) among
(c) between (d) each

Set-3 Fill in the blanks

1. Distribute these books ________ the boys.
(a) between (b) amongs
(c) among (d) amid

2. She came and sat ________ me.
(a) besides (b) beside
(c) by (d) under

3. I met Meera ________ gave me your phone number.
(a) which (b) whom
(c) who (d) whose

4. She does not smoke nor ________ her husband.
(a) did (b) do
(c) does (d) could

5. I ________ happy since she came in this city.
(a) am (b) was
(c) have been (d) used to

6. Open rebuke is ________ than secret love.
(a) best (b) warmer
(c) better (d) well

7. I am used to ________ tea.
(a) take (b) takes
(c) taking (d) took

8. If I ________ you, I would help you.
(a) am (b) was
(c) were (d) have been

9. I made the horse ________ run.
(a) to (b) by
(c) with (d) none of these

10. You must read the ________ of this novel carefully.
(a) content (b) contents
(c) contentment (d) syllabus.

11. Translate this passage into Hindi ________.
(a) word to word (b) word by word
(c) word for word (d) none of these

12. You need not ________write him an application.
(a) to (b) upon
(c) over (d) none of these

13. The examination will commence ________ 25 March.
(a) on (b) from
(c) by (d) at

14. This is the same book ________ you gave me yesterday.
(a) who (b) whose
(c) that (d) which

15. Nobody helped me yesterday, ________?
(a) did he (b) does he
(c) did she (d) did they

16. I gave him a five hundred rupee not to buy ________.
(a) few book (b) a few book
(c) a few books (d) the few books

17. It is the third time she ________ to office with them.
(a) had gone (b) has gone
(c) went (d) goes

18. One of my friends ________ knowledge of phonetics.
(a) having (b) have
(c) has (d) none of these

19. She has lived in Delhi ________ the last ten years.
(a) since (b) from
(c) for (d) by

20. Please do open your book ________ page number 56.
(a) on (b) at
(c) upon (d) above

21. It will have been raining ________ 5 O'clock.
(a) since (b) for
(c) over (d) by me

22. Cut your coat according your ________.
(a) mean (b) means
(c) cloth (d) legs

23. John complained to the professor ________ Aliston.
(a) with (b) against
(c) on (d) none of these

24. She has no house to line ________.
(a) in (b) with
(c) on (d) under

25. It is ________ who am going to help you.
(a) I (b) my
(c) me (d) I

Set-4 Fill in the blanks

1. The police are ________ on his activities.
(a) keeping an eye (b) keeping watchful eyes
(c) keeping both eyes (d) keeping one eye

2. Are you ________ stamp collecting?
(a) interested by (b) interested in
(c) interested at (d) interested about

3. Your report ________ my statement.
(a) conforms (b) conforming
(c) confirms (d) conforting

4. He has no recommendations, he will ________ get the job.
(a) however (b) what so ever
(c) where as (d) whatever

5. Leaves ________ in autumn.
(a) fall down (b) fall of
(c) fall on (d) fall off

6. Each time he tried to climb the tree; he failed to do ________.
(a) thus (b) than
(c) so (d) this

7. The man was ________ his deafness.
(a) rewarded for (b) cured of
(c) convicted by (d) saved at

8. He is the talk ________.
(a) of the town (b) for the town
(c) about the town (d) over the town.

9. I ________ very much like to see my old classmates.
(a) can (b) should
(c) would (d) might

10. A partial victory is sometimes all we can hope to ________.
(a) gain (b) lose
(c) suffer (d) reward

11. My father was advised to ________ his smoking.
(a) cut down (b) cut off
(c) cut into (d) cut in

12. I had not expected to meet him; it was quite an ________ meeting.
(a) organised (b) unique
(c) accidental (d) intentional

13. Pour the juice ________ the glass.
(a) in (b) into
(c) onto (d) on

14. I have been ________ this problem since noon.
(a) working into (b) working on
(c) working for (d) with after

15. Sushma has a deep ________ for drunkards.
(a) hatred (b) hate
(c) hated (d) hating

16. Nothing is impossible ________ him.
(a) in (b) about
(c) for (d) on

17. My relations ________ him are not friendly.
(a) between (b) with
(c) to (d) for

18. George ________ to his country a year ago.
(a) was coming (b) came
(c) is coming (d) comes

19. One should always set ________ some money for emergency.
(a) aside (b) on
(c) by (d) out

20. The train is expected to arrive in a few minutes ________ platform six.
(a) in (b) above
(c) near (d) at

21. I have not had ________ opportunity to study during the day.
(a) very (b) many
(c) much (d) more

22. We can ________ see anything in the dim light.
(a) badly (b) scarcely
(c) fairly (d) rather

23. Few countries can ________ India in variety, colour and the richness of dance forms.
(a) compare (b) rival
(c) perform (d) prevail

24. The scheme allows students from different countries to communicate ________.
(a) each other (b) with each other
(c) themselves (d) with one another

25. Rita promised to come but she never turned ________.
(a) in (b) out
(c) down (d) up

Set-5 Fill in the blanks

1. I cannot conceive of ________ accepting the post.
(a) he (b) his
(c) him (d) he will

2. The maestro's son has no taste ________ music.
(a) about (b) with
(c) in (d) for

3. The starving man ________ able to walk.
(a) can barely (b) were barely
(c) are barely (d) was barely

4. The new railway counter at Dwarka should ________ all travellers.
(a) benefit (b) reward
(c) reserve (d) provide

5. My younger brother ________ my grandfather.
(a) looks on (b) looks after
(c) looks to (d) looks of

6. Due to heavy rains, the match was called ________.
(a) of (b) off
(c) down (d) out

7. The management gave us permission to ________ with the new curriculum.
(a) to fixed (b) go about
(c) go ahead (d) go around

8. Discrimination _____ any form should be avoided
(a) of (b) by
(c) from (d) in

9. I like ________ in the sun.
(a) laying (b) lay
(c) lie (d) lying

10. Please do not ________ an offer made by the chairman.
(a) deny (b) refuse
(c) refrain (d) refute

11. The clerk asked for my ________.
(a) determination (b) destiny
(c) destination (d) designation

12. This is the wrong ________ to this question.
(a) saying (b) style
(c) reaction (d) answer

13. The resort is ________ the lake.
(a) from (b) has
(c) besides (d) beside

14. The life of mosquito is ________ but very active.
(a) long (b) passive
(c) brief (d) important

15. The director congratulated Mr. Verma ________ his success.
(a) on (b) for
(c) at (d) about

16. The boys were ________ to hear that we were going to build a bridge.
(a) delight (b) delights
(c) delighted (d) delighting

17. They have a cup of tea ________?
(a) do they (b) don't they
(c) did they (d) don't we

18. The revolting players and the game's administrators hold ________ discussions to resolve the payment dispute.
(a) unnecessary (b) obligatory
(c) preliminary (d) silent

19. I hope you know that, once you have signed the contract you will not be able to ________.
(a) back in (b) back out
(c) back up (d) back at

20. Your opening paragraph should ________ the reader's attention.
(a) atleast (b) address
(c) attract (d) affect

21. When I visited him last evening we talked the matter ________.
(a) through (b) away
(c) off (d) over

22. Most children remain ________ school between the ages of seven and ten.
(a) in (b) under
(c) at (d) inside

23. The quake also ________ mudslides on me outskirts of the city.
(a) tempered (b) tampered
(c) erupted (d) triggered

24. The king was moved with ________ for the beggar.
(a) pity (b) piety
(c) pithy (d) privy

25. The doctor has been ________ to attend to an urgent case.
(a) called (b) called in
(c) called by (d) called for

Set-6 Fill in the blanks

1. He ran so fast ________.
(a) that he fell (b) so he fell
(c) he fell (d) that he is falling

2. The gate ________ by the watchman on duty.
(a) opened (b) was opened
(c) was opened (d) was opening

3. The king was ________ while he was addressing his subjects.
(a) murdered (b) executed
(c) assassinated (d) exterminated

4. Her parents will never give their ________ to such an unsuitable match.
(a) acquiescence (b) consent
(c) agreement (d) willingness

5. The authorities would not ________ to the striker's demands.
(a) accept (b) accede
(c) accession (d) access

6. With the amount of conflicts in the world, a timed world war is ________.
(a) emanate (b) emanant
(c) eminent (d) imminent

7. The traditional way of saying that you feel fine is to say that you are ________.
(a) alright (b) all correct
(c) all good (d) all right

8. The army has been ________ to be ready to control the floods.
(a) alerted (b) warned
(c) awakened (d) wanted

9. Never ________ met any girl so beautiful.
(a) have I (b) I have
(c) I had (d) did I

10. He went there ________.
(a) with foot (b) by foot
(c) walking on foot (d) on foot

11. After your long illness I am happy to see you ________.
(a) by and by (b) up and about
(c) up and down (d) up and round

12. Making pies and cakes ________ Mrs. Reddy's speciality.
(a) is (b) has
(c) were (d) are

13. The deserted garden was infested ________ weeds.
(a) into (b) with
(c) on (d) for

14. You must ________ your career with all seriousness.
(a) pursue (b) complete
(c) follow (d) direct

15. He set up institutions of international ________.
(a) famous (b) reputation
(c) renown (d) repute

16. If you do well in the exam next year then my efforts ________ successful.
(a) will be (b) have been
(c) would have been (d) has been

17. The noise of the drumbeats ________ and frightened the tiger.
(a) pertained (b) persisted
(c) prevented (d) perplexed

18. We ________ not hurry, we have got plenty of time.
(a) would (b) must
(c) should (d) need

19. Masked man ________ a security van on the motorway.
(a) held out (b) hold forth
(c) held on (d) help up

20. I ________ reading this book since morning.
(a) has been (b) have been
(c) was (d) am

21. We should ________ the rules.
(a) cope with (b) comply with
(c) conduce to (d) side with

22. Coward die many times ________ their death.
(a) about (b) before
(c) after (d) for

23. Give me a cheque _____ Rs. 10,000
(a) in (b) for
(c) of (d) with

24. The new discovery is ________ phenomenal.
(a) novelly (b) beautifully
(c) always (d) indeed

25. You take a decision the ball is in ________ now.
(a) your court (b) your net
(c) your garden (d) your packet

Set-7 Fill in the blanks

1. I could hardly recognise him ________ I saw him.
(a) after (b) but
(c) and (d) when

2. A garden knife is ________ used for light pruning.
(a) systematically (b) daily
(c) generally (d) compulsorily

3. My India by Corbett deals ________ the author's familiarity with and love of India.
(a) at (b) with
(c) in (d) of

4. There are not solitary, free living creatures; every form of life is ________ other forms.
(a) overshadowed by (b) segregated from
(c) dependent on (d) parallel to

5. Serious threat to out ecology and environment can be ________ with organic cultivation.
(a) aggravated (b) combated
(c) hastened (d) impeded

6. I'll take ________ now as I have another appointment somewhere else.
(a) permission (b) leave from work
(c) departure (d) my leave

7. We had a ________ deadline.
(a) tight (b) easy
(c) difficult (d) quick

8. She has great affection ________ the body
(a) at (b) in
(c) for (d) with

9. I am accustomed to ________ tea with one of my friends at this tea stall.
(a) taking (b) takes
(c) take (d) took

10. This ring is made ________ gold.
(a) from (b) with
(c) of (d) by

11. Let you and ________ go there to meet him.
(a) I (b) my
(c) mine (d) me

12. Either you or I ________ reading this book.
(a) am (b) were
(c) are (d) does

13. Pecola is as ________ as Radha.
(a) wisest (b) mere wise
(c) wise (d) wiser

14. ________ few books that you gave me, were interesting.
(a) a (b) an
(c) the (d) none of these

15. In my opinion, she ________ to be a good teacher.
(a) seeming (b) seemed
(c) seems (d) seem

16. They had reached the railway station before the train ________.
(a) had left (b) had been left
(c) left (d) was leaving

17. He slipped ________ his old ways and started drinking again.
(a) into (b) off
(c) by (d) in

18. All is not well ________ the automobile sector.
(a) of (b) down
(c) in (d) to

19. See me tomorrow without ________.
(a) fails (b) failure
(c) fail (d) failing

20. Radium ________ rays that blind the eyes.
(a) gives away (b) gives over
(c) gives (d) gives off

21. You will have to work ________ you are in the office.
(a) as long as (b) as good as
(c) as far as (d) as much as

22. Either you ________ your brother has to come tomorrow.
(a) nor (b) and
(c) as well as (d) or

23. You as well as I ________ composing a sonnet since morning.
(a) have been (b) are
(c) is (d) am

24. The houses were covered ________ the snow.
(a) by (b) with
(c) over (d) alongwith

25. He was ________ in the river without making any attempt.
(a) swimming (b) floating
(c) jumping (d) none of these.

Set-8 Fill in the blanks

1. It is cloudy, the rain ________.
(a) eminent (b) imminent
(c) averse to (d) is imminent

2. There is no key to ________ these questions without practice.
(a) solves (b) solving
(c) solved (d) solution

3. I want to transfer this cheque ________ her name.
(a) by (b) in
(c) on (d) with

4. The doctor advised him to eat ________ food.
(a) healthy (b) healthful
(c) clean (d) scruffy

5. Let us go there, ________?
(a) will you (b) can you
(c) shall I (d) shall we

6. He loves reading, writing and ________ the poor in this spare time.
(a) helps (b) to helping
(c) helping (d) to help

7. The slum children ________ round a burning fire to keep themselves warm.
(a) grouped (b) huddled
(c) tucked (d) elbowed

8. Polyster shirts are more ________ than the cotton ones.
(a) durably (b) duration
(c) durability (d) durable

9. She was born ________ rich parents.
(a) of (b) from
(c) with (d) to

10. Children should obey ________ parents.
(a) they (b) their
(c) mine (d) them

11. The child has been messing ________ yesterday.
(a) for (b) by
(c) to (d) since

12. I do not like ________ sugar in my coffee.
(a) many (b) much
(c) few (d) fewer

13. The board results are ________ any moment.
(a) expected (b) launched
(c) compiled (d) tabulated

14. Nobody likes ________ coming late in the classroom.
(a) they (b) them
(c) theirs (d) their

15. It is high time you ________ an elegy.
(a) composing (b) composes
(c) composed (d) compose

16. It was ________ who were making a noise.
(a) them (b) they
(c) their (d) theirs

17. Sheela is averse to ________ the poor.
(a) help (b) helping
(c) helps (d) helped

18. This novel is ________ better of the two
(a) a (b) an
(c) the (d) none of these

19. The gentry of the town ________ been invited by the president.
(a) is (b) was
(c) has (d) have

20. I doubt ________ she can help me in this case.
(a) that (b) if
(c) because (d) so that

21. The road was lined ________ the people.
(a) by (b) with
(c) at (d) none of these

22. The teacher asked me what my name ________.
(a) has been (b) is
(c) were (d) was

23. Why did you hurt the ________ of my right foot?
(a) finger (b) fingers
(c) toe (d) none of these

24. Seldom ________ he come to attend the class on time.
(a) do (b) does
(c) is (d) are

25. Pecola questioned me ________ I know me way to station.
(a) if (b) whether
(c) that (d) weather

Answer Key : Set-1 Fill in the blanks

1. (c) 2. (a) 3. (b) 4. (d) 5. (c) 6. (b)
7. (c) 8. (c) 9. (c) 10. (c) 11. (c) 12. (b)
13. (c) 14. (c) 15. (d) 16. (c) 17. (c) 18. (c)
19. (a) 20. (d) 21. (c) 22. (c) 23. (c) 24. (d)
25. (b)

Answer Key : Set-2 Fill in the blanks

1. (d) 2. (d) 3. (d) 4. (d) 5. (d) 6. (a)
7. (b) 8. (a) 9. (c) 10. (c) 11. (a) 12. (a)
13. (b) 14. (b) 15. (d) 16. (b) 17. (a) 18. (b)
19. (d) 20. (c) 21. (a) 22. (c) 23. (c) 24. (a)
25. (d)

Answer Key : Set-3 Fill in the blanks

1. (c) 2. (b) 3. (c) 4. (c) 5. (c) 6. (c)
7. (c) 8. (c) 9. (d) 10. (b) 11. (c) 12. (d)
13. (a) 14. (c) 15. (d) 16. (c) 17. (b) 18. (c)
19. (c) 20. (b) 21. (d) 22. (c) 23. (b) 24. (a)
25. (a)

Answer Key : Set-4 Fill in the blanks

1. (a) 2. (b) 3. (c) 4. (a) 5. (d) 6. (c)
7. (b) 8. (a) 9. (c) 10. (a) 11. (a) 12. (c)
13. (b) 14. (b) 15. (a) 16. (c) 17. (b) 18. (b)
19. (a) 20. (d) 21. (d) 22. (b) 23. (d) 24. (d)
25. (d)

Answer Key : Set-5 Fill in the blanks

1. (b) 2. (d) 3. (d) 4. (a) 5. (b) 6. (b)
7. (c) 8. (d) 9. (d) 10. (b) 11. (d) 12. (d)
13. (d) 14. (c) 15. (a) 16. (c) 17. (b) 18. (b)
19. (d) 20. (c) 21. (b) 22. (a) 23. (c) 24. (b)
25. (d)

Answer Key : Set-6 Fill in the blanks

1. (a) 2. (c) 3. (c) 4. (b) 5. (a) 6. (d)
7. (d) 8. (a) 9. (a) 10. (d) 11. (b) 12. (a)
13. (b) 14. (a) 15. (d) 16. (a) 17. (d) 18. (d)
19. (d) 20. (b) 21. (b) 22. (b) 23. (b) 24. (c)
25. (a)

Answer Key : Set-7 Fill in the blanks

1. (d) 2. (c) 3. (b) 4. (d) 5. (a) 6. (b)
7. (a) 8. (c) 9. (a) 10. (c) 11. (d) 12. (a)
13. (c) 14. (c) 15. (c) 16. (a) 17. (a) 18. (c)
19. (c) 20. (d) 21. (a) 22. (d) 23. (a) 24. (b)
25. (b)

Answer Key : Set-8 Fill in the blanks

1. (d) 2. (b) 3. (b) 4. (b) 5. (d) 6. (c)
7. (a) 8. (d) 9. (a) 10. (b) 11. (d) 12. (b)
13. (a) 14. (d) 15. (c) 16. (b) 17. (b) 18. (c)
19. (d) 20. (a) 21. (b) 22. (d) 23. (c) 24. (b)
25. (d)

CHAPTER 16

Sentence Improvement

Set-1 Sentence Improvement

1. I know him for ten years.
   (a) knew him (b) has known him
   (c) have known him (d) no improvement

2. Open the door, will you?
   (a) won't you (b) can't you
   (c) don't you (d) no improvement

3. It was the third time, he went to school with Raman.
   (a) goes (b) has gone
   (c) had gone (d) no improvement

4. She has written this book last year.
   (a) wrote (b) writes
   (c) writing (d) no improvement

5. Work hard, lest you may fail.
   (a) should not (b) should
   (c) could (d) no improvement

6. Hardly he had reached the station when it began to rain.
   (a) Hardly did he (b) Hardly had he
   (c) scarcely he had (d) no improvement

7. When I went to school, I take my lunch.
   (a) I took my lunch.
   (b) I have taken my lunch.
   (c) I want to take my lunch.
   (d) No improvement

8. The man that I met yesterday was dancing on the road.
   (a) who (b) which
   (c) whom (d) no improvement

9. I am afraid you two are at cross purpose.
   (a) are at cross purposes
   (b) are in cross purposes
   (c) are with cross purposes
   (d) no improvement

10. Can you tell me where has he gone?
    (a) where has gone he
    (b) where gone has he
    (c) where he has gone
    (d) no improvement

11. No sooner I heard the shot when I rushed to the sport.
    (a) No sooner I heard the short then
    (b) No sooner did I hear the shot when
    (c) No sooner did I hear the shot than
    (d) no improvement

12. The teacher said that a man is a social animal.
    (a) a man is a social animal
    (b) the man is social animal
    (c) man is a social animal
    (d) no improvement

13. The most birds were seen flying in the sky.
    (a) the most bird (b) most bird
    (c) most birds (d) no improvement

14. Having wrote a letter, he posted it.
    (a) Have written (b) Having writes
    (c) Having written (d) no improvement

15. Seeing the lion, he ran away.
    (a) to see the lion
    (b) having seen the lion
    (c) to be seeing the lion
    (d) no improvement

16. How long do you love her?
    (a) does you love her
    (b) have you been loving her
    (c) have you loved her
    (d) no improvement

17. Raman talks as if he knows everything about America.
    (a) he knew (b) he has known
    (c) he has known (d) no improvement

18. I am happy, since he came here.
(a) I was happy (b) I have happy
(c) I have been happy (d) no improvement

19. The master was good at using pleasant names for unpleasant things in order to hoodwink the labourer.
(a) euphemism (b) euphoria
(c) pleasantries (d) no improvement

20. He gave me so many informations regarding this case.
(a) so much informations
(b) so many information
(c) so many pieces of information
(d) no improvement

21. There is no place for you to sit at this desk.
(a) no space (b) no room
(c) no vacant place (d) no improvement

22. They have brought a new pen for their exams.
(a) a new pens (b) much pens
(c) some new pens (d) no improvement

23. This is the only time she cooks food.
(a) she cooked food
(b) she is cooking food
(c) she has cooked food
(d) no improvement

24. My father will be operated tomorrow.
(a) would be operate
(b) would be operated
(c) will be operated upon
(d) no improvement

25. The teacher advised them to cooperate each other if they want to progress.
(a) cooperate one another
(b) cooperate each other
(c) cooperate
(d) no improvement

Set-2 Sentence Improvement

1. She has slept for eight hours last night.
(a) slept (b) had slept
(c) has been sleeping (d) no improvement

2. I have seen the film and she also has.
(a) has also (b) has too
(c) too has (d) no improvement

3. I hope that I shall get a first class.
(a) I feel that (b) I hope
(c) I am hoping (d) no improvement

4. You can mix with some sugar and eat it.
(a) in some sugar (b) into some sugar
(c) any sugar (d) no improvement

5. He stopped to work an hour ago.
(a) to working (b) to have worked
(c) working (d) no improvement

6. The problem of translation are still remain.
(a) are remain (b) will remained
(c) will still remain (d) no improvement

7. It is high time you come to a decision
(a) came (b) had come
(c) have come (d) no improvement

8. I have bought this house in 1970 for Rs. two lakh.
(a) had bought (b) bought
(c) have been bought (d) no improvement

9. I have seen much of the plays of Shakespeare.
(a) a few (b) many
(c) most (d) no improvement

10. The child died from jaundice.
(a) with (b) of
(c) by (d) no improvement

11. If you have studied hard, you would have got a first class.
(a) If you studied hard
(b) If you had studied hard
(c) If you would study hard
(d) no improvement

12. She did not like the movie, or I did.
(a) nor I died (b) nor did I
(c) nor do I (d) no improvement

13. Will you me visit and come in the hospital.
(a) come and visit me
(b) visit me and come
(c) come and me visit
(d) no improvement

14. Your hair needs cutting badly.
(a) Badly your hair needs cutting
(b) Your hair needs cutting real bad
(c) Your hair badly needs cutting
(d) no improvement

15. If only he can swim.
(a) would (b) could
(c) will (d) no improvement

16. Please examine my homework.
(a) see (b) check
(c) investigate (d) no improvement

17. To tell you the truth, I am not there last night.
(a) is (b) was
(c) were (d) no improvement

18. It will have been raining since morning.
(a) by (b) by the
(c) since the (d) no improvement

19. Who does teach you English?
(a) do teach (b) does teaches
(c) teaches (d) no improvement

20. She asked me where my father was.
(a) is (b) were
(c) has been (d) no improvement

21. Milk is contained by this bottle.
(a) on (b) in
(c) with (d) no improvement

22. Have you ever gone to Holland?
(a) want (b) goes
(c) been (d) no improvement

23. This is the best book which you gave me.
(a) whose (b) of which
(c) that (d) no improvement

24. I wanted to purchase a T.V. last year.
(a) a T.Vs (b) T.V.
(c) a T.V. set (d) no improvement

25. Dr. Faustus wrote his name with blood.
(a) by blood (b) in blood
(c) on blood (d) no improvement

Set-3 Sentence Improvement

1. Her new shoes ensure her that she does not fall while running.
(a) ensures to her that (b) ensures her that
(c) ensure that (d) no improvement

2. I am not understanding your problem.
(a) I am understand (b) I do not understand
(c) I understand (d) No improvement

3. She returned the computer after she noticed it was damaged.
(a) return the computer after
(b) returned a computer after
(c) return a computer then
(d) no improvement

4. Return this book to me when you read it
(a) when you are reading it.
(b) when you had read it.
(c) when you have read it.
(d) no improvement

5. Mohan and not has relatives are coming.
(a) were (b) is
(c) am (d) no improvement

6. I want to purchase a toothpaste.
(a) an toothpaste (b) toothpastes
(c) a tube of toothpaste(d) no improvement

7. I have been telling her that she had better consulted a good doctor.
(a) consulting (b) consult
(c) been consulting (d) no improvement

8. He is as miser as Roman.
(a) as mise as (b) as miserly as
(c) so mise as (d) no improvement

9. The number of crimes have been increasing rapidly.
(a) has been increasing
(b) are being increasing
(c) have increased
(d) no improvement

10. I have read three fourth of this book.
(a) three and fourth (b) three fourthes
(c) three fourths (d) no improvement

11. You must listen my words carefully.
(a) listening (b) listen to
(c) to listen (d) no improvement

12. Sandra has good command on mathematics.
(a) command over (b) command upon
(c) command of (d) no improvement

13. Reading of this book will help you a lot in your exams.
(a) To read of this book
(b) The reading of this book
(c) Reading of the book
(d) no improvement

14. You must complete this work up to Sunday.
(a) within Sunday (b) by Sunday
(c) on to Sunday (d) no improvement

15. She has no pen to write.
(a) write with (b) write at
(c) write on (d) no improvement

16. Hole wheat bread is good for health.
(a) whole (b) healthy
(c) holed (d) no improvement

17. He is one of those who likes to help others.
(a) Those persons who likes to help others
(b) Those who like to help others
(c) Those who like to help other
(d) no improvement

18. He is entitled to a reward for honesty.
(a) titled (b) entitle to
(c) entitled (d) no improvement

19. The sight chosen for the school building is good.
(a) site (b) eye sight
(c) side (d) no improvement

20. The matter must be considered in every point of view.
(a) from every (b) at every
(c) on every (d) no improvement

21. The poor villagers have waited in the bitter cold for more than two hours now.
(a) has been waiting (b) had waited
(c) have been waiting (d) no improvement

22. The car with a burst tyre halted swiftly.
(a) stopped quick (b) stopped fastly
(c) Jolted to a halt (d) no improvement

23. Many incidents of the last month seem unimportant when viewed in perception.
(a) prospective (b) perceptive
(c) perspective (d) no improvement

24. Will you land me few rupees in this hour of need.
(a) borrow me a few rupees
(b) lend me any rupees
(c) lend me a few rupees
(d) no improvement

25. You must accustom yourself with new ideas.
(a) accustomed with (b) accustom to
(c) accustom yourself (d) no improvement

Set-4 Sentence Improvement

1. When I saw him, he wrote an essay.
(a) he had written an essay
(b) he was written an essay
(c) he was writing an essay
(d) no improvement

2. Somebody has watered the plants, hasn't they?
(a) don't they (b) have they
(c) haven't they (d) no improvement

3. It is the third time. I saw him smoking.
(a) I see him smoke
(b) I saw him smoke
(c) I have seen him smoking
(d) no improvement

4. Raja not only passed and won a scholarship.
(a) not passed but won
(b) not only passed but also won
(c) not only passed and also won
(d) no improvement

5. He is averse to drink wine.
(a) drank wine (b) drinks wine
(c) drinking wine (d) no improvement

6. Completed my homework, I went out to play.
(a) The work was finished
(b) Homework completed
(c) Having completed my homework
(d) no improvement

7. My friend, along with his cousins have gone to watch a movie at the nearest cinema hall.
(a) has been going (b) has gone
(c) all going (d) no improvement

8. It is about time we went home.
(a) go (b) goes
(c) gone (d) no improvement

9. I gave him a few coins that I had in my pocket.
(a) few coins (b) the few coin
(c) the few coins (d) no improvement

10. The question is the easiest of the two.
(a) easy (b) easier
(c) the easier (d) no improvement

11. Being Sunday, I did not go to office.
(a) Having Sunday (b) It Sunday
(c) It being Sunday (d) no improvement

12. You should have thoroughly knowledge of grammar.
(a) soundly (b) through
(c) nicely (d) no improvement

13. It's a great movie, you might go and see it. You will really like it.
(a) should go (b) could go
(c) would go (d) no improvement

14. I would soon walk than stand in the cold waiting for the bus.
(a) hardly (b) longer
(c) sooner (d) no improvement

15. He has given to gambling and drinking.
(a) given away (b) given in
(c) given up (d) no improvement

16. Ramu never makes at least a century each time he plays cricket.
(a) seldom (b) forever
(c) always (d) no improvement

17. The tribe lived with different customs.
(a) on (b) by
(c) through (d) no improvement

18. The illness means, he must resign the chairmanship impromptu.
(a) fast (b) forthwith
(c) earliest (d) no improvement

19. The judge gave the verdict against the government.
(a) pronounced (b) declared
(c) sentenced (d) no improvement

20. The fire service personnel put off the fire with much difficultly.
(a) put on (b) put out
(c) put away (d) no improvement

21. He has the irritating habit of playing his own trumpet all the time.
(a) blowing his own trumpet
(b) pumping his own trumpet
(c) bringing up his own trumpet
(d) no improvement

22. All people want to be happy do they?
(a) don't they (b) are they
(c) didn't they (d) no improvement

23. Salman and the other actors drive through the forest.
(a) in (b) over
(c) among (d) no improvement

24. My friend has a toothache and cannot eat something.
(a) anything (b) things
(c) edibles (d) no improvement

25. Either Lata or Mala have done it.
(a) Neither Lata or Mala has done it.
(b) Either Lata or Mala has done it.
(c) Neither Lata or Mala have done it.
(d) no improvement

Set-5 Sentence Improvement

1. Africa is in the South of Europe.
(a) by the (b) to the
(c) ever the (d) no improvement

2. He enjoys to go on tours with the children.
(a) to go for tours (b) going for tours
(c) going on tours (d) no improvement

3. Sita was cooking in the kitchen when her husband was ringing the bell.
(a) would have been (b) ringing
(c) rang (d) no improvement

4. Never such incidents have taken place on our campus.
(a) have such incidents
(b) such incidents will have
(c) were incidents
(d) no improvement

5. My mother teach me to be a good human being.
(a) taught (b) thought
(c) talked (d) no improvement

6. Will you tell me who did accompany you to the airport?
(a) when did accompany you
(b) who must be accompanied you
(c) who accompanied you
(d) no improvement

7. Navinder and his wife departed for London by Air India yesterday.
(a) to (b) towards
(c) toward (d) no improvement

8. The child was too small to learn English alphabets.
(a) English alphabet (b) Alphabets of English
(c) English's alphabets(d) no improvement

9. I have this car since 2015.
(a) I am having this car(b) I had this car
(c) I have had this car (d) no improvement

10. I and my parents live in Delhi.
(a) Me and my parents (b) My parents and I
(c) My parents and me (d) no improvement

11. Somebody will show you the way to station tomorrow.
(a) will shows you the way
(b) Will has shown the way
(c) must be show you the way
(d) no improvement

12. The alarming report of the plane crash left everyone in a state of shock.
(a) alarmed report (b) alarmed reporting
(c) reported alarm (d) no improvement

13. He is wiser than handsome.
(a) the wiser (b) the wisest
(c) more wise (d) no improvement

14. Meera and myself will help them.
(a) Myself and Meera (b) Meera and me
(c) Meera and I (d) no improvement

15. He would surely assist me if I had requested him to do so.
(a) would surely assisted me
(b) would have surely assisted me
(c) will have surely assisted me
(d) no improvement

16. Knowing that the area was prone to earthquakes all the buildings were reinforced with additional concrete.
(a) On being prone to earthquakes
(b) Having been knowing that the area was prone to earthquakes
(c) Since the area was known to be prone to earthquakes
(d) no improvement

17. No sooner did she finish the lecture, that the students began to leave
(a) then the students began to leave
(b) than the students began to leave
(c) and the students began to leave
(d) no improvement

18. Jumping up, he ran away
(a) she runs away
(b) she is running away
(c) she has run away
(d) no improvement

19. You are advised to fill this form with black ink.
(a) in (b) by
(c) from (d) no improvement

20. It is me who will show the way to you
(a) It is I (b) It is my
(c) It is mine (d) no improvement

21. A black and white cow are grazing
(a) A black and white cows are
(b) The black and white cow is
(c) A black and white cow is
(d) no improvement

22. I am going to give my exams tomorrow.
(a) going to take
(b) going to took
(c) going to giving
(d) no improvement

23. The greater the demand, higher the price
(a) the high (b) a higher
(c) the higher (d) no improvement

24. He did not know how to solve the problem and I did not neither.
(a) neither do I (b) neither did I
(c) either did I (d) no improvement

25. They entered into my room at 10:30 yesterday.
(a) enter in (b) entered on
(c) entered (d) no improvement

Set-6 Sentence Improvement

1. Supposing if he comes, what should I do?
(a) If he comes (b) If he come
(c) If he came (d) no improvement

2. I prefer to ride than to walk
(a) ride to walk
(b) riding than walking
(c) riding to walking
(d) no improvement

3. His speech was broadcasted over the radio.
(a) were broadcasted
(b) was broadcast
(c) have been broadcast
(d) no improvement

4. If I was you, I would not sign the document
(a) If I had been you
(b) If I were you,
(c) If I have been you
(d) no improvement

5. Kalidas is the Shakespear of India.
(a) Shakespeare (b) like Shakespeare
(c) a Shakespeare (d) no improvement

6. No sooner had she heard the news when the fainted.
(a) that she fainted (b) than she fainted
(c) and she fainted (d) no improvement

7. I saw him laughing on the mendicant.
(a) laugh on (b) laughing at
(c) laughed at (d) no improvement

8. He is the only one of those writers who have written many books.
(a) writer who has (b) writers who have
(c) writers that has (d) no improvement

9. He is enough kind to help you.
(a) kind enough (b) kinder enough
(c) too enough kind (d) no improvement

10. I wish I have a car.
(a) I am having (b) I must have
(c) I had (d) no improvement

11. She is feeling comparatively better now.
(a) comparatively good
(b) comparatively best
(c) comparatively well
(d) no improvement

12. The tea cup that I bought yesterday for my son were stole.
(a) had been (b) have been
(c) was (d) no improvement

13. Would you mind me opening the door?
(a) me open
(b) my opening
(c) mine opening
(d) no improvement

14. One must obey his parents.
(a) their parents (b) her parents
(c) one's parents (d) no improvement

15. You should help the poors and the needy.
(a) poor and needy
(b) the poor and needy
(c) the poor and the needy
(d) no improvement

16. Her four hairs seems to be gray.
(a) four hair seem (b) four hair seems
(c) four hairs seem (d) no improvement

17. Despite of working hard she could not get even pass marks.
(a) Despite (b) Despite being
(c) Inspite (d) no improvement

18. Could you discuss about this matter tomorrow.
(a) discuss on (b) discuss
(c) discuss over (d) no improvement

19. Who do not know the name of B.R. Ambedkar?
(a) Who did not know
(b) Who do not know
(c) Who does not know
(d) no improvement

20. Would you mind help me with these questions?
(a) to help (b) of helping
(c) helping (d) no improvement

21. I went to home with a view to meeting Ram yesterday.
(a) go to home (b) goes to home
(c) went home (d) no improvement

22. They have written this play in 2016.
(a) had written
(b) wrote
(c) have had written
(d) no improvement

23. I am not able to make out the meaning for your statement.
(a) the meaning of your statement
(b) the meaning and your statement
(c) the meaning from your statement
(d) no improvement

24. Cindy had applied for the post of an editor.
(a) the editor (b) editor
(c) some editor (d) no improvement

25. Balinda came crying in fall of 1994.
(a) on fall of 1994 (b) in the fall of 1994
(c) over fall of 1994 (d) no improvement

Set-7 Sentence Improvement

1. She insisted to go there, though her husband cautioned her on it.
(a) on going (b) upon going
(c) going (d) no improvement

2. He asked me if go to school yesterday.
(a) if I have gone to school yesterday,
(b) if I went to the school yesterday.
(c) if I went to school the previous day
(d) no improvement

3. Hema could not prepare tea because there were a little milk in the pot.
(a) there were little milk
(b) there was a little milk
(c) there was little milk
(d) no improvement

4. The audience show the clown performing.
(a) saw the clown performing
(b) saw the clown to perform
(c) saw the clown perform
(d) no improvement

5. The monkey was sitting on the tree.
(a) in (b) upon
(c) over (d) no improvement

6. The student told the teacher to explain the passage.
(a) told to the teacher (b) said to the teacher
(c) asked the teacher (d) no improvement

7. Ram cooked a conspiracy to cheat Abdul.
(a) cultivated (b) hatched
(c) fabricated (d) no improvement

8. You could have gave me a five hundred rupee note to buy a few books.
(a) could has given (b) could have given
(c) could have give (d) no improvement

9. The burglar broke in the shop last night.
(a) broke (b) broke through
(c) broke into (d) no improvement

10. Old habits die hardly.
(a) Old habits dies hard
(b) Old habits die hard
(c) Older habits die too hard
(d) no improvement

11. All the members of his family are in Ernakulam.
(a) All the family members of his
(b) All his family members
(c) All of his family members
(d) no improvement

12. Being tired he sat down under the tree.
(a) He being tired (b) Being tired
(c) Having tired (d) no improvement

13. He cut vegetables with a knife, does he?
(a) doesn't he (b) did he
(c) didn't he (d) no improvement

14. I did not see you for a long time.
(a) saw you
(b) have not seen you
(c) have not seen
(d) no improvement

15. If I have time, I will ring you up.
(a) I would (b) I could
(c) I must (d) no improvement

16. They have not and cannot be in the good books oft he coach, because they lack discipline.
(a) have not and can never been
(b) have not and can never be
(c) have not been and can never be
(d) no improvement

17. I want to admit in a university in the US.
(a) enroll (b) enter
(c) go (d) no improvement

18. My words must be listened by you.
(a) listen to by you (b) listened to with you
(c) listened to by you (d) no improvement

19. Many a man have died of cancer.
(a) has died of
(b) have died from
(c) have been dying of
(d) no improvement

20. A hand pump is very easy to work and it can be fitted in every house.
(a) can be fits (b) can be fitting
(c) could be fitting (d) no improvement

21. She went to market for purchasing a note book.
(a) for buying (b) to purchase
(c) to purchased (d) no improvement

22. You made the children to cry.
(a) cry (b) cried
(c) crying (d) no improvement

23. There is nothing which I should tell you
(a) whom (b) who
(c) that (d) no improvement

24. The criminal was hung to death.
(a) hanged (b) hang
(c) hunged (d) no improvement

25. Wave after wave surrounded the tower.
(a) engulfed (b) circled
(c) encircled (d) no improvement

Set-8 Sentence Improvement

1. Grandma is not feeling well and want to lay down.
(a) lain (b) laid
(c) lie (d) no improvement

2. My doctor adviced me to go to the gym more often.
(a) advice (b) advise
(c) advised (d) no improvement

3. The service trip to control America had a life-altering affect on Rosemary.
(a) effect on (b) affects on
(c) effects (d) no improvement

4. To dance good, you have to practice a lot.
(a) Dancing good (b) To dancing well
(c) To dance well (d) no improvement

5. Bobby has been laying his clothes on the bed instead of folding them.
(a) laid (b) lying
(c) lain (d) no improvement

6. Tina is looking for a pet who is small and easy to care for.
(a) a pet whom is (b) a pet that are
(c) a pet that is (d) no improvement

7. It is him who will be responsible for making all the arrangements.
(a) his (b) him
(c) he (d) no improvement

8. It is he who often provides me a classical pizza on Sundays.
(a) classic pizza (b) a classic pizza
(c) classical pizza (d) no improvement

9. If you see anyone on the trail, tell them to be careful.
(a) tells them (b) tell him
(c) tells him (d) no improvement

10. Nobody brought their book despite my request
(a) his book
(b) our book
(c) its book
(d) no improvement

11. Elliot lives in Chicago for ten years.
(a) live in Chicago
(b) is living in Chicago
(c) has lived in Chicago
(d) no improvement

12. The deer we saw was tall, alert and it was obviously hungry.
(a) and it is obviously hungry
(b) and obviously hungry
(c) and was obvious hungry
(d) no improvement

13. Unless you do not come on time, I will not teach you.
(a) Unless you came
(b) Unless you come
(c) Unless you had come
(d) no improvement

14. John was so feeling bad that he couldn't go to school and had to go to the doctor instead.
(a) was feeling so bad (b) was feeling too bad
(c) felt so bad enough (d) no improvement

15. Candida was born in poor parents.
(a) born from
(b) born off
(c) born of
(d) no improvement

16. Three people from work had to be going to prison because they had stealed money from the company.
(a) were having gone to prison because they stealed
(b) went to the prison because they had stolen
(c) went to prison because they had stolen
(d) no improvement

17. They are going to Shimla with a view to meet their relatives.
(a) to met
(b) with a view to met
(c) with a view to meeting
(d) no improvement

18. Ravi felt at home between his classmates at the school reunion.
(a) amid (b) amongst
(c) among (d) no improvement

19. My little brother cries as a baby.
(a) cry as a baby (b) cries as baby
(c) cries like a baby (d) no improvement

20. I have been working on the car engine since this morning.
(a) I am working (b) I used to working
(c) I am to work (d) no improvement

21. By whom the Paradise lost written?
(a) Who are Paradise lost written?
(b) By whom was the Paradise lost written?
(c) To whom was the Paradise lost written?
(d) no improvement

22. Each of the class members have one vote.
(a) having one vote (b) has vote
(c) has one vote (d) no improvement

23. Ask him to take this medicine when ate food.
(a) then he eats food
(b) when he eats food
(c) when he has eaten food
(d) no improvement

24. My house is three times bigger than yours.
(a) three times big than
(b) three times as bigger as
(c) three times as big as
(d) no improvement

25. He wrote this book before she returned from Japan.
(a) He had written this book
(b) He could have written this book
(c) He was writing this book
(d) no improvement

Answer Key : Set-1 Sentence Improvement

1. (c)	2. (d)	3. (b)	4. (a)	5. (b)	6. (b)
7. (a)	8. (c)	9. (d)	10. (c)	11. (c)	12. (c)
13. (c)	14. (a)	15. (d)	16. (c)	17. (a)	18. (c)
19. (a)	20. (c)	21. (b)	22. (c)	23. (c)	24. (c)
25. (c)					

Answer Key : Set-2 Sentence Improvement

1. (d)	2. (a)	3. (d)	4. (d)	5. (c)	6. (c)
7. (a)	8. (b)	9. (b)	10. (b)	11. (b)	12. (b)
13. (a)	14. (d)	15. (b)	16. (b)	17. (b)	18. (b)
19. (c)	20. (d)	21. (b)	22. (c)	23. (c)	24. (c)
25. (b)					

Answer Key : Set-3 Sentence Improvement

1. (d)	2. (b)	3. (d)	4. (c)	5. (b)	6. (c)
7. (b)	8. (b)	9. (a)	10. (c)	11. (b)	12. (c)
13. (b)	14. (b)	15. (a)	16. (a)	17. (b)	18. (d)
19. (a)	20. (a)	21. (c)	22. (d)	23. (c)	24. (c)
25. (b)					

Answer Key : Set-4 Sentence Improvement

1. (c)	2. (c)	3. (c)	4. (b)	5. (c)	6. (c)
7. (b)	8. (d)	9. (c)	10. (c)	11. (c)	12. (b)
13. (a)	14. (c)	15. (c)	16. (c)	17. (b)	18. (c)
19. (d)	20. (d)	21. (a)	22. (a)	23. (d)	24. (a)
25. (b)					

Answer Key : Set-5 Sentence Improvement

1. (b)	2. (c)	3. (c)	4. (a)	5. (b)	6. (c)
7. (d)	8. (a)	9. (c)	10. (b)	11. (d)	12. (d)
13. (c)	14. (c)	15. (b)	16. (d)	17. (b)	18. (d)
19. (a)	20. (a)	21. (c)	22. (a)	23. (c)	24. (b)
25. (c)					

Answer Key : Set-6 Sentence Improvement

1. (a)	2. (c)	3. (b)	4. (b)	5. (d)	6. (b)
7. (d)	8. (c)	9. (a)	10. (c)	11. (a)	12. (c)
13. (b)	14. (c)	15. (c)	16. (c)	17. (a)	18. (b)
19. (c)	20. (c)	21. (c)	22. (b)	23. (a)	24. (b)
25. (b)					

Answer Key : Set-7 Sentence Improvement

1. (a)	2. (c)	3. (c)	4. (c)	5. (a)	6. (c)
7. (b)	8. (b)	9. (c)	10. (b)	11. (d)	12. (d)
13. (c)	14. (b)	15. (d)	16. (c)	17. (a)	18. (c)
19. (a)	20. (d)	21. (b)	22. (a)	23. (c)	24. (a)
25. (d)					

Answer Key : Set-8 Sentence Improvement

1. (d)	2. (c)	3. (a)	4. (c)	5. (d)	6. (c)
7. (c)	8. (b)	9. (b)	10. (a)	11. (c)	12. (b)
13. (b)	14. (a)	15. (c)	16. (c)	17. (c)	18. (a)
19. (c)	20. (d)	21. (b)	22. (c)	23. (c)	24. (c)
25. (a)					

CHAPTER 17

Common Errors

COMMON ERRORS Set-1

1. One must learn (a) to distinguish (b) good from bad. (c) No error (d)

2. The children (a) laughed at (b) the clown. (c) No error (d)

3. Had the plane not been delayed (a) I will reach there (b) in time for the function. (c) No error (d)

4. In India, hill stations (a) usually have (b) beautiful sceneries. (c) No error (d)

5. She was ill for five days (a) when the doctor (b) was sent for. (c) No error (d)

6. He is not (a) as coward as (b) his brother. (c) No error (d)

7. As he is honest (a) and he is (b) liked by all. (c) No error (d)

8. I am not sure (a) if I have told you already (b) but I have decided to get a new job. (c) No error (d)

9. Our teacher told us (a) that the earth (b) moved around the sun. (c) No error (d)

10. Raja Ram Mohan Roy was greatly pained (a) when he heard the cried of widows (b) being burnt on their husband's pyres. (c) No error (d)

11. Neither of these dresses (a) give you (b) that great look. (c) No error (d)

12. If I went to Agra (a) | I would have seen (b) | the Taj Mahal. (c) | No error (d)

13. It is I (a) | who is (b) | going to Kanpur. (c) | No error (d)

14. The book that (a) | you gave me (b) | yesterday was very good. (c) | No error (d)

15. They have (a) | this house (b) | since 2010. (c) | No error (d)

16. These tests (a) | show that someone (b) | is having an infection. (c) | No error (d)

17. No sooner (a) | I had started for college (b) | than it began to rain (c) | No error (d)

18. Either he or his friends (a) | is managing (b) | this show. (c) | No error (d)

19. Rajat's brother as well as I (a) | am guilty (b) | of cheating Ramesh. (c) | No error (d)

20. Taking regular exercises (a) | are good (b) | for health. (c) | No error (d)

21. A fire was started accidently (a) | but we were able (b) | to put it out. (c) | No error (d)

22. I bought (a) | this book (b) | in Rs. 200. (c) | No error (d)

23. Romeo went (a) | to school (b) | to meet his friend (c) | No error (d)

24. She asked me (a) | why were they (b) | making a noise. (c) | No error (d)

25. I am happy (a) | since (b) | he came here. (c) | No error (d)

COMMON ERRORS Set-2

1. It is high time (a) / she goes (b) / to office. (c) / No error (d)

2. One should analyses (a) / his mistakes (b) / if one wants to progress. (c) / No error (d)

3. Scarcely did I reach the airport nervous and tense (a) / than the plane took off (b) / leaving stranded in the alien place. (c) / No error (d)

4. Nothing would prevent (a) / him to speaking (b) / out against injustice (c) / No error (d)

5. The reason why (a) / we were late is because (b) / there was much traffic. (c) / No error (d)

6. The principal forbade us (a) / not to use (b) / the entire hall. (c) / No error (d)

7. There is no room (a) / at this desk (b) / for you. (c) / No error (d)

8. There is no key to (a) / learn English (b) / without proper practice. (c) / No error (d)

9. A group (a) / of people were (b) / standing at the bus stop. (c) / No error (d)

10. She had been (a) / teaching in this school (b) / since the last ten years. (c) / No error (d)

11. Nisha is one of those girls (a) / who lives (b) / in this city for five years. (c) / No error (d)

12. Two people started (a) / this project (b) / and only one person remained. (c) / No error (d)

13. I have read through the complaint (a) / once so I am not (b) / ready to sign on it . (c) / No error (d)

14. He questioned me (a) | if I knew (b) | the way to the station. (c) | No error (d)

15. The dog moved (a) | it's puppies (b) | everytime we came near. (c) | No error (d)

16. My dog barks (a) | when it (b) | see my neighbour's cat. (c) | No error (d)

17. The puppies (a) | in the dog house (b) | is adorable. (c) | No error (d)

18. I haven't felt good (a) | since (b) | the weather changed. (c) | No error (d)

19. This is the (a) | worse meal (b) | I have ever had. (c) | No error (d)

20. My collection of stamps (a) | include many (b) | from around the world. (c) | No error (d)

21. Me arriving late (a) | throw the schedule (b) | off completely. (c) | No error (d)

22. John and myself (a) | usually solve (b) | the problems easily. (c) | No error (d)

23. I like classical (a) | music much more (b) | than she does. (c) | No error (d)

24. The match will be (a) | telecasted on DD1 (b) | tomorrow. (c) | No error (d)

25. Nobody comes (a) | on time (b) | does he? (c) | No error (d)

COMMON ERRORS Set-3

1. The police (a) | has arrested him (b) | on the charge of theft. (c) | No error (d)

2. The period (a) | between 1991 to 1996 (b) | was very significant in my life. (c) | No error (d)

3. Chritina's friends (a) | preferred her remedies (b) | over the doctor's. (c) | No error (d)

4. Those who are in power (a) | have to be sensitive of (b) | the suffering of the poor. (c) | No error (d)

5. You will come (a) | to my sister's wedding tomorrow (b) | isn't it? (c) | No error (d)

6. My elder brother (a) | asked me (b) | that what I was doing. (c) | No error (d)

7. The Hindu is (a) | most popular than any other (b) | newspaper in India. (c) | No error (d)

8. We should arrange (a) | for a porter (b) | as the luggage are heavy. (c) | No error (d)

9. Grapes (a) | cannot gathered (b) | from thistles. (c) | No error (d)

10. He went to Mumbai (a) | with a view (b) | to secure a job. (c) | No error (d)

11. The headmaster with all his senior (a) | teachers have (b) | come to attend the meeting. (c) | No error (d)

12. The redevelopment project is aimed (a) | not just providing good houses to shanty dwellers (b) | but also developing infrastructure around the major Mumbai localists. (c) | No error (d)

13. The poet (a) | describes about (b) | the spring season (c) | No error (d)

14. Due to me being a new comer (a) | I was unable (b) | to get a good house (c) | No error (d)

15. The social activist (a) | was murder (b) | in cold blood. (c) | No error (d)

16. A city dweller finds it difficult (a) / to pass away the time (b) / in a village. (c) / No error (d)

17. The fifth and final act (a) / of Macbeth contain (b) / the sleep walking scene. (c) / No error (d)

18. My sister (a) / has read (b) / pages after pages of the Bible. (c) / No error (d)

19. Beside Hindi (a) / she knew (b) / Bengali well. (c) / No error (d)

20. The actress (a) / was shocked (b) / by the news of her dog's death. (c) / No error (d)

21. John is working very hardly (a) / as the examinations (b) / are fast approaching. (c) / No error (d)

22. He ate (a) / nothing (b) / since yesterday. (c) / No error (d)

23. The fireman wore (a) / inflammable clothing (b) / for protection. (c) / No error (d)

24. Visitors (a) / were not permitted to (b) / entering the park after darkness. (c) / No error (d)

25. Adults suffering chicken pox (a) / can develop (b) / all kinds of complications. (c) / No error (d)

COMMON ERRORS Set-4

1. The well known Pianoist (a) / had to practise for several hours a day (b) / even after he rose to fame. (c) / No error (d)

2. A hot and a cold spring (a) / was found (b) / near each other. (c) / No error (d)

3. All doubts (a) | are cleared (b) | between you and I. (c) | No error (d)

4. Either of the roads (a) | lead (b) | to the park. (c) | No error (d)

5. I have passed (a) | the examination (b) | two years ago. (c) | No error (d)

6. He said (a) | that he will never (b) | repeat the mistake. (c) | No error (d)

7. The passer by told us (a) | where was the marriage hall (b) | and even led us to it. (c) | No error (d)

8. The increase in consumption is directly (a) | proportional to the increase (b) | in income. (c) | No error (d)

9. The officer (a) | is angry on the clerk (b) | for not attending to the work. (c) | No error (d)

10. The knowledge of (a) | atleast two languages (b) | are required to pass the examination. (c) | No error (d)

11. She did no (a) | compose an elegy, (b) | does she? (c) | No error (d)

12. Being (a) | a rainy day (b) | I could not go out. (c) | No error (d)

13. When you have gone through the papers (a) | kindly return them (b) | to us. (c) | No error (d)

14. Ten kilometers (a) | is (b) | a long distance to walk. (c) | No error (d)

15. He asked (a) | supposing if he fails (b) | what he would do. (c) | No error (d)

16. I could not put at a hotel (a) | because the board and loading charges (b) | were too expensive. (c) | No error (d)

17. There is no agreement (a) / between you and I (b) / so we are free to go our way. (c) / No error (d)

18. It is necessary for everything to prepare himself (a) / thoroughly in his early years for the tasks (b) / that he will have to perform in life. (c) / No error (d)

19. Those who lay down their lives (a) / for the sake of others will surely (b) / dwell forever in a world of bliss. (c) / No error (d)

20. Many a man (a) / have come to India from Bangladesh (b) / to live here permanently. (c) / No error (d)

21. The first inning (a) / of the match (b) / was very sensational. (c) / No error (d)

22. On a holiday (a) / I prefer reading story books (b) / than visiting my friends. (c) / No error (d)

23. Seldom if ever (a) / nature does operate (b) / in closed and separate compartments. (c) / No error (d)

24. Neither of the girls (a) / were willing to (b) / accept the proposal. (c) / No error (d)

25. According to me (a) / he is (b) / a good man (c) / No error (d)

COMMON ERRORS Set-5

1. The students (a) / were playing (b) / besides the school. (c) / No error (d)

2. Mango the most unique fruit (a) / is available in India (b) / in plenty. (c) / No error (d)

3. When I was passing through the forest (a) I happened to see (b) a number of deers (c) No error (d)

4. He is guilty (a) for killing (b) an innocent bird. (c) No error (d)

5. Every boy and every girl (a) were given (b) a pocket of sweets. (c) No error (d)

6. The number of guests (a) invited to the wedding (b) are one thousand. (c) No error (d)

7. I did not receive (a) any letter from my parents (b) for the last two months. (c) No error (d)

8. It is said (a) that this room is not being opened (b) for the last fifty years. (c) No error (d)

9. There are so many filths (a) all around (b) the place. (c) No error (d)

10. My cousin with her parents (a) have gone (b) to Chennai. (c) No error (d)

11. He left (a) his home town (b) with bag and baggage last year (c) No error (d)

12. I have just come (a) to know that Mr. Ray one of my favourite (b) teachers, died with cancer recently. (c) No error (d)

13. Rabindranath Tagore was not (a) the first Nobal Laureate (b) without a university degree. (c) No error (d)

14. The company has ordered (a) some (b) new equipments. (c) No error (d)

15. She is one of the (a) | best mothers (b) | that has ever lived. (c) | No error (d)

16. Neither the mouse (a) | nor the lion (b) | were caught. (c) | No error (d)

17. Judge in him (a) | prevailed upon the father (b) | and he sentenced his son to death. (c) | No error (d)

18. A cup of coffee (a) | is an excellent complement (b) | to smoked salmon. (c) | No error (d)

19. One major reason (a) | for the popularity of television is (b) | that most people like to stay at home. (c) | No error (d)

20. The servant (a) | hanged up the lamp (b) | on the wall. (c) | No error (d)

21. The curator told us (a) | the museum was not opened (b) | to the public on Sunday. (c) | No error (d)

22. The Railways have made (a) | crossing the tracks (b) | a punished offence. (c) | No error (d)

23. The short story (a) | should not exceed (b) | more than two hundred words. (c) | No error (d)

24. I know(a) | a doctor (b) | you are reffering to. (c) | No error (d)

25. Although the police officer sympathised with poor, (a) | he refused to (b) | take an action against the rich man. (c) | No error (d)

Answer Key ➤ Common Errors : SET-1

1. (c)	2. (d)	3. (b)	4. (c)	5. (a)	6. (b)
7. (b)	8. (d)	9. (c)	10. (b)	11. (b)	12. (b)
13. (b)	14. (d)	15. (a)	16. (c)	17. (b)	18. (b)
19. (b)	20. (b)	21. (a)	22. (c)	23. (b)	24. (b)
25. (a)					

Explanations ➤ Common Errors : SET-1

1. (c)
Replace *good from bad* by between good and bad.

2. (d)
No error.

3. (b)
Replace *I will there* by I would have reached there.

4. (c)
Replace *beautiful sceneries* by beautiful scenery.

5. (a)
Replace *she was ill for five days* by she had been ill for fine days.

6. (b)
Replace *as coward as* by as cowardly as.

7. (b)
Replace and he is by he is.
NOTE : as is not followed by and.

8. (d)
No error.

9. (d)
Replace *moved around the Sun* by moves around the Sun.

Note : Universal truth requires ***Present Simple Tense***.

10. (b)
Replace *cried of widows* by cries of widows.

11. (b)
Replace *give you* by gives you.

12. (b)
Replace *I would have seen* by I would see.

13. (b)
Replace *who is* by who am.

14. (d)
No error.

15. (a)
Replace *they have* by they have had.

16. (c)
Replace *is having an infection* by someone has an infection.

17. (b)
Replace *I had started for college* by had I started for college.

18. (b)
Replace *is managing* by are managing.

19. (b)
Replace *an guilty* by is guilty.

20. (b)
Replace *are good* by is good.

21. (a)
Replace *a fire was started accidentally* by a fire started accidentally.

22. (c)
Replace *in Rs 200* by for Rs 200.

23. (b)
Replace *to school* by to the school.

24. (b)
Replace *why were they* by why they were.

25. (a)
Replace *I am happy* by I have been happy.

Answer Key ➤ Common Errors : SET-2

1. (b)	2. (b)	3. (b)	4. (b)	5. (b)	6. (b)
7. (d)	8. (b)	9. (b)	10. (c)	11. (b)	12. (c)
13. (c)	14. (b)	15. (b)	16. (c)	17. (c)	18. (a)
19. (b)	20. (b)	21. (b)	22. (a)	23. (d)	24. (b)
25. (c)					

Explanations ➤ Common Errors : SET-2

1. (b)
Replace *she goes* by she went.

2. (b)
Replace *his mistakes* by one's mistake.

3. (b)
Replace *than the plane took off* by when the plane took off.

4. **(b)**
Replace *him to speaking* by him from speaking.

5. **(b)**
Replace *because* by that.

6. **(b)**
Replace *not to use* by to use.

7. **(d)**
No error.

8. **(b)**
Replace *learn English* by learning English.

9. **(b)**
Replace *of people were* by of people was.

10. **(c)**
Replace *since the last ten years* by for the last ten year.

11. **(b)**
Replace *who lives* by who have lived.

12. **(c)**
Replace *and only one person remained* by but only one person remained.

13. **(c)**
Replace *ready to sign on it* by ready to sign it.

14. **(b)**
Replace *if I knew* by whether I knew.

15. **(b)**
Replace *it's puppies* by its puppies.

16. **(c)**
Replace *see my neighbour's cat* by sees my neighbour's cat.

17. **(c)**
Replace *is adorable* by are adorable.

18. **(a)**
Replace *I haven't felt good* by I haven't felt well.

19. **(b)**
Replace *worse meal* by worst meal.

20. **(b)**
Replace *include many* by includes many.

21. **(b)**
Replace *throw the schedule* by throws the schedule.

22. **(a)**
Replace *John myself* by John and I.

23. **(d)**
No error.

24. **(b)**
Replace *telecasted on DD1* by telecast on DD1.

25. **(c)**
Replace *does he?* by doesn't he?.

Answer Key ➤ Common Errors : SET-3

1. (b)	2. (b)	3. (c)	4. (b)	5. (c)	6. (c)
7. (b)	8. (c)	9. (b)	10. (c)	11. (b)	12. (b)
13. (b)	14. (a)	15. (b)	16. (b)	17. (b)	18. (c)
19. (a)	20. (c)	21. (a)	22. (a)	23. (b)	24. (c)
25. (a)					

Explanations ➤ Common Errors : SET-3

1. **(b)**
Replace *has arrested him* by have arrested him.

2. **(b)**
Replace *between 1991 to 1996* by between 1991 and 1996.

3. **(c)**
Replace *over the doctor's* by to the doctor's.

4. **(b)**
Replace *have to be sensitive of* by have to be sensitive to.

5. **(c)**
Replace *isn't it?* by won't you?.

6. **(c)**
Replace *that what I was doing* by what I was doing.

7. **(b)**
Replace *most popular than any* by more popular than any.

8. **(c)**
Replace *as the luggage are heavy* by as the luggage is heavy.

9. **(b)**
Replace *cannot gathered* by cannot be gathered.

10. **(c)**
Replace *to secure a job* by securing a job.

11. **(b)**
Replace *teachers have* by teachers has.

12. **(b)**
Replace *not just providing* by not only providing.

13. **(b)**
Remove about.
Note : ***Describe*** *is not followed* by any ***preposition***.

14. **(a)**
Replace *due to me being a new comer* by due to my being a new comer.

15. **(b)**
Replace *was murder* by was murdered.

16. **(b)**
Replace *to pass away the time* by to pass the time.

17. **(b)**
Replace *of Macbeth contain* by of Macbeth contains.

18. **(c)**
Replace *pages after pages of the Bible* by page after page of the Bible.

19. **(a)**
Replace *beside Hindi* by besides Hindi.

20. **(c)**
Replace *by the mews of dog's death* by at the news of her dog's death.

21. **(a)**
Replace *John is working very hardly* by John is working very hard.

22. **(a)**
Replace *he ate* by he had eaten.

23. **(b)**
Replace *inflammable clothing* by not inflammable clothing.

24. **(c)**
Replace *entering the park after darkness* by to enter the park after darkness.

25. **(a)**
Replace *adults suffering chicken pox* by adults suffering from chicken pox.

Answer Key ➤ Common Errors : SET-4

1. (d)	2. (b)	3. (c)	4. (b)	5. (a)	6. (b)
7. (b)	8. (d)	9. (b)	10. (c)	11. (c)	12. (a)
13. (d)	14. (d)	15. (b)	16. (b)	17. (b)	18. (a)
19. (d)	20. (b)	21. (d)	22. (c)	23. (b)	24. (b)
25. (a)					

Explanations ➤ Common Errors : SET-4

1. **(d)**
No error.

2. **(b)**
Replace *was found* by were found.

3. **(c)**
Replace *between you and I* by between you and I me.

4. **(b)**
Replace *lead* by leads.

5. **(a)**
Replace *I have passed* by I passed.

6. **(b)**
Replace *that he will never* by that he would never.

7. **(b)**
Replace *where was the marriage hall* by where the marriage hall was.

8. **(d)**
No error.

9. **(b)**
Replace *is angry on the clerk* by is angry with the clerk.

10. **(c)**
Replace *are required to pass* by is required to pass.

11. **(c)**
Replace *does she* by did she?

12. **(a)**
Replace *being* by It being.

13. **(d)**
No error.

14. **(d)**
No error.

15. **(b)**
Supposing and if cannot be used together, so remove either if or supposing.

16. **(b)**
Replace *because the board and* by because the boarding and.

17. **(b)**
Replace *between you and I* by between you and me.

18. **(d)**
No error.

19. (d)
No error.

20. (b)
Replace *have come to India* by has come to India.

21. (d)
No error.

22. (c)
Replace *than visiting my friends* by to visiting my friends.

23. (b)
Replace *nature does operate* by does nature operate.

24. (b)
Replace *were willing to* by was willing to.

25. (a)
Replace *according to me* by in my opinion.
Note : ***me*** **and** ***us*** **cannot be used** ***after according to*****.**

Answer Key ➤ Common Errors : SET-5

1. (c)	2. (a)	3. (c)	4. (b)	5. (b)	6. (c)
7. (a)	8. (b)	9. (a)	10. (b)	11. (c)	12. (c)
13. (d)	14. (c)	15. (c)	16. (c)	17. (a)	18. (d)
19. (d)	20. (b)	21. (c)	22. (c)	23. (c)	24. (b)
25. (a)					

Explanations ➤ Common Errors : SET-5

1. (c)
Replace *besides the school* by beside the school.

2. (a)
Replace *mango the most unique fruit* by mango a unique fruit.

3. (c)
Replace *a number of deers* by a number of deer.

4. (b)
Replace *for killing* by of killing.

5. (b)
Replace *were given* by was given.

6. (c)
Replace *are one thousand* by is one thousand.

7. (a)
Replace *I did not receive* by I had not received.

8. (b)
Replace *that this room is not being opened* by that this room has not been opened.

9. (a)
Replace *there are so many filths* by there is so much filth.

10. (b)
Replace *have gone* by has gone.

11. (c)
Do not use with before bag and baggage.

12. (c)
Replace *teachers died with* ..... by teachers died of.

13. (d)
No error.

14. (c)
Replace *new equipments* by new equipment.

15. (c)
Replace *that has ever lived* by that have ever lived.

16. (c)
Replace *were caught* by was caught.

17. (a)
Replace *Judge in him* by the Judge in hum.

18. (d)
No error.

19. (d)
No error.

20. (b)
Replace *hanged up the lamp* by hung up the lamp.

21. (c)
Replace *to the public on Sunday* by for the public on Sunday.

22. (c)
Replace *a punished offence* by a punishable offence.

23. (c)
Replace *more than two hundred words* by two hundred words.

24. (b)
Replace *a doctor* by the doctor.

25. (a)
Replace .... *with poor* by .... the poor.

CHAPTER 18

Analogy

Analogy

An analogy is a similarity in some respects between things that are otherwise dissimilar. In each of the questions given in this section, you will find a pair of related words. These words are separated by a colon.

EXAMPLES —

Shield : Arrow

(a) Bow : Quiver

(b) Knight : Horse

(c) Umbrella : Rain drops

(d) Boy : Girl

EXPLANATIONS —

A ***shield*** protects from ***arrows***; an ***umbrella*** protects from ***rain drops***.

The best way to solve an '***analogy question***' is to make up a sentence that describes the relationship between the first two words and find another pair in the choices that would fit into that same sentence.

PRACTICE EXERCISE Analogy

Direction : Complete the second pair based on the relation shown in the first pair.

1. Fossils : Creatures : : Mummies : ?

(a) Egypt (b) Human beings
(c) Animals (d) Martyrs

2. Moon : Satellite : : Earth : ?

(a) Sun (b) Planet
(c) Solar system (d) Asteroid

3. Clock : Time : : Thermometer : ?

(a) Heat (b) Radiation
(c) Energy (d) Temperature

4. Paw : Cat : : Hoof : ?

(a) Horse (b) Lion
(c) Lamb (d) Elephant

5. Fruits : Basket : : Fishes : ?

(a) Rip (b) River
(c) Room (d) Cup

Direction : Each of the following questions consists of two words that have a certain relationship to each other, followed by four options. Select the pair which indicates the same relationship as the original pairs of words.

6. Restaurant : Menu

(a) Library : Catalogue

(b) Artist : Imitative

(c) Book : Encyclopaedia

(d) Guest : Rude

7. Calligraphy : Writing
(a) Music : Song (b) Lyric : Poem
(c) Drama : Prose (d) Chapter : Stanza

8. Chocolate : Sugar
(a) Egg : Yolk (b) Road : Traffic
(c) Building : Cement (d) Milk : Cream

9. Fish : Shoal
(a) Wolf : Pack (b) Tiger : Jungle
(c) Herd : Peacock (d) Dog : Collie

10. Scales : Weight
(a) Yardstick : Length
(b) Width : Depth
(c) Size : Area
(d) Mileage : Speed

11. Starving : Hungry
(a) Neat : Thoughtful
(b) Towering : Cringing
(c) Progressive : Regressive
(d) Depressed : Sad

12. Professor : Erudite
(a) Aviator : Licensed
(b) Inventor : Imaginative
(c) Procrastinator : Conscientious
(d) Moderator : Vicious

13. Spy : Clandestine
(a) Accountant : Meticulous
(b) Furrier : Rambunctious
(c) Lawyer : Ironic
(d) Astronaut : Opulent

14. Hat : Head
(a) Cold : Hot (b) Winter : Snow
(c) Clothes : Closet (d) Glove : Hand

15. Metaphor : Symbol
(a) Pentameter : Poem
(b) Rhythm : Melody
(c) Analogy : Comparison
(d) Slang : Usage

16. Teacher : School
(a) Actor : Role
(b) Judge : Courthouse
(c) Author : Book
(d) Mechanic : Engine

17. Search : Find
(a) Sleep : Wake
(b) Explore : Discover
(c) Draw : Paint
(d) Walk : Run

18. Phobic : Fearful
(a) Ridiculous : Silly
(b) Shy : Familiar
(c) Envious : Desiring
(d) Cautious : Emotional

19. Slapstick : Laughter
(a) Fallacy : Dismay
(b) Genre : Mystery
(c) Horror : Fear
(d) Mimicry : Tears

20. Dirge : Funeral
(a) Chain : Letter
(b) Bell : Church
(c) Telephone : Call
(d) Jingle : Commercial

Answer Key > Analogy

1. (b)	2. (b)	3. (d)	4. (a)	5. (b)	6. (a)
7. (b)	8. (c)	9. (a)	10. (a)	11. (d)	12. (b)
13. (a)	14. (d)	15. (c)	16. (b)	17. (b)	18. (a)
19. (c)	20. (d)				

Explanations > Analogy

1. (b)
Fossils are the remains of creatures in the same way mummies are remains of human beings.

2. (b)
Moon is a satellite, in the same way Earth is a planet.

3. (d)
Clock tells time, in the same way thermometer tells temperature.

4. (a)
Paw belongs to cat, in the same way hoof belongs to horse.

5. (b)

Second is a container to carry the first.

6. (a)

Restaurant has a menu in the same way library has a catalogue.

7. (b)

Calligraphy is an art of good hand writing in the same way lyric is an art of a good poem.

8. (c)

Chocolate is made with the help of sugar in the same way building is built with the help of cement.

9. (a)

A group of fish is called a shoal, and a group of wolves is called a pack.

10. (a)

Scales are used to measure weight in the same way a yardstick is used to measure length.

11. (d)

Starving is an intensification of hungry in the same way depressed is an intensification of sad.

12. (b)

Being erudite is a trait of a professor in the same way, being imaginative is a trait of an inventor.

13. (a)

A spy acts in a clandestine manner in the same way an accountant acts in a meticulous manner.

14. (d)

A hat is worn on the head in the same way a glove is worn on the hand.

15. (c)

A metaphor is a symbol, and an analogy is a comparison.

16. (b)

A teacher works in a school, and a Judge works in a courthouse.

17. (b)

One searches in order to find, and in the same way one explores in order to discover.

18. (a)

To be phobic is to be extremely fearful, and to be ridiculous is to be extremely silly.

19. (c)

Slapstick results in laughter, in the same way horror results in fear.

20. (d)

A dirge is a song used at funeral, and a jingle is a song used in a commercial.

Section B

VOCABULARY

CHAPTER 1

Enhance Your Word Power

Enhance Your Word Power		
Sl.	Word	Meaning
1.	Penury (Noun)	Extreme poverty
2.	Petulant (Adjective)	Rude, Peevish
3.	Quagmire (Noun)	A difficult situation
4.	Perturb (Verb)	To disturb
5.	Carping (Verb)	To find fault
6.	Chasten (Verb)	To correct by punishment
7.	Stygian (Adjective)	Dark and gloomy
8.	Tirade (Noun)	A long violent speech
9.	Rue (Verb)	To regret
10.	Venerate (Verb)	To respect, honor, adore
11.	Fulsome (Adjective)	Abundant
12.	Imbroglio (Noun)	An extremely confused or complicated situation
13.	Toothsome (Adjective)	Delicious, sexually attractive
14.	Insipid (Adjective)	Tasteless
15.	Futile (Adjective)	Completely useless
16.	Ubiquitous (Adjective)	Omnipresent, found everywhere
17.	Flag (Verb)	To become week, feeble

Enhance Your Word Power		
Sl.	Word	Meaning
18.	Compromise (Verb)	To reduce the quality
19.	Coin (Verb)	To invent a word or phrase
20.	Dearth (Noun)	Paucity, Shortage of something
21.	Ravenous (Adjective)	Very hungry
22.	Subtle (Adjective)	Difficult to detect
23.	Clandestine (Adjective)	Secret, not open
24.	Indigenous (Adjective)	Native to an area
25.	Enigmatic (Adjective)	Mysterious
26.	Obfuscate (Verb)	To confuse, Perplex
27.	Abjure (Verb)	To reject
28.	Abrogate (Verb)	To abolish, To finish
29.	Alacrity (Noun)	Eagerness
30.	Antipathy (Noun)	Strong dislike
31.	Blandish (Verb)	To coax by flattery
32.	Conflagration (Noun)	Great fire
33.	Contrite (Verb)	To remorse, to feel sorry
34.	Cupidity (Noun)	Greed, strong desire
35.	Decry (Verb)	To criticise, Excoriate

Enhance Your Word Power		
Sl.	Word	Meaning
36.	Enervate (Verb)	To weaken
37.	Evince (Verb)	To show, reveal
38.	Ebullient (Adjective)	Extremely lively
39.	Approbation (Noun)	Praise
40.	Myriad (Adjective)	Many, numerous
41.	Ameliorate (Verb)	To improve, increase
42.	Lucid (Adjective)	Clear, easily understood
43.	Ornate (Verb)	To decorate
44.	Eloquent (Adjective)	Well spoken, expressive
45.	Exonerate (Verb)	To remove blame
46.	Foment (Verb)	To stir up, incite
47.	Gregarious (Adjective)	Sociable, outgoing
48.	Inert (Adjective)	Lethargic, unmoving, sluggish
49.	Fleece (Verb)	To deceive, swindle, dupe
50.	Wane (Verb)	To decrease
51.	Thrifty (Adjective)	A person who spends money wisely or using money wisely
52.	Penurious (Adjective)	Miserly
53.	Affluent (Adjective)	Wealthy
54.	Ruminate (Verb)	To think, ponder
55.	Immaculate (Adjective)	Spotless, Clean, pure

Enhance Your Word Power		
Sl.	Word	Meaning
56.	Sordid (Adjective)	Filthy, dirty
57.	Render (Verb)	To give, deliver
58.	Rattle (Verb)	To make a noise
59.	Advocate (Verb)	To support
60.	Felony (Noun)	A major crime
61.	Congenial (Adjective)	Pleasant, friendly
62.	Accord (Verb)	To agree
63.	Flaunt (Verb)	To display proudly
64.	Abstruse (Adjective)	Hard to understand
65.	Diffident (Adjective)	Shy, timid
66.	Dour (Adjective)	Harsh
67.	Indolent (Adjective)	Lazy, Lethargic
68.	Noxious (Adjective)	Harmful, injurious
69.	Sensitise (Verb)	To make aware
70.	Wrangle (Verb)	To argue
71.	Rampant (Adjective)	Unchecked, something that cannot be controlled
72.	Feign (Verb)	To pretend
73.	Lucrative (Adjective)	Profitable
74.	Elegant (Adjective)	Attractive
75.	Profound (Adjective)	Deep, great
76.	Proclaim (Verb)	To announce something
77.	Serene (Adjective)	Calm and peaceful

Enhance Your Word Power		
Sl.	Word	Meaning
78.	Refurbish (Verb)	To decorate
79.	Extol (Verb)	To praise someone
80.	Rebuke (Verb)	To scold, chide, reprimand, berate
81.	Garrulous (Adjective)	Talkative
82.	Repudiate (Verb)	To reject
83.	Exhume (Verb)	To remove from a grave
84.	Nostalgia (Noun)	Homesickness
85.	Pensive (Adjective)	Thoughtful in a sad way
86.	Doleful (Adjective)	Full of sorrow, mournful
87.	Dejected (Adjective)	Sad, in law spirits, depressed
88.	Elated (Adjective)	Joyful
89.	Complacent (Adjective)	Self satisfied
90.	Buxom (Adjective)	Plump and attractive
91.	Benign (Adjective)	Not dangerous gentle
92.	Salubrious (Adjective)	Healthful
93.	Zenith (Noun)	The highest point, culmination
94.	Nadir (Noun)	The lowest point
95.	Longevity (Noun)	Long life
96.	Erudite (Adjective)	Skilled, scholarly, learned

Enhance Your Word Power		
Sl.	Word	Meaning
97.	Esoteric (Adjective)	Hard to understand
98.	Euphoria (Noun)	Feeling of great happiness
99.	Heyday (Noun)	A time of great success
100.	Relegate (Verb)	To transfer to a lower position
101.	Bolster (Verb)	To support
102.	Cognition (Noun)	Mental ability to know things
103.	Incognito (Adjective)	Under a false identity
104.	Conspicuous (Adjective)	Easily seen
105.	Blather (Verb)	To talk nonsense
106.	Queer (Adjective)	Strange
107.	Envisage (Verb)	To imagine, to visualize
108.	Uncanny (Adjective)	Unusual, strange
109.	Crestfallen (Adjective)	Disappointed
110.	Abate (Verb)	To reduce, assuage
111.	Affinity (Noun)	Similarity, Kinship
112.	Aggrandise (Verb)	To improve, enhance
113.	Appellation (Noun)	Name
114.	Amalgamate (Verb)	To mix
115.	Ardent (Adjective)	Passionate
116.	Belligerent (Adjective)	Quarrelsome

Enhance Your Word Power		
Sl.	**Word**	**Meaning**
117.	Assiduous (Adjective)	Hard working
118.	Apprehensive (Adjective)	Worried
119.	Turpitude (Noun)	Sin
120.	Vandalize (Verb)	To damage public or private property
121.	Humane (Adjective)	Kind hearted
122.	Inhume (Verb)	To burry
123.	Inundate (Verb)	To flood
124.	Burgeon (Verb)	To expand, to flourish
125.	Bequest (Noun)	Something left to someone in a will
126.	Cajole (Verb)	To persuade someone to do something
127.	Candor (Noun)	Truthfulness
128.	Capricious (Adjective)	Likely to change at any time
129.	Circumspect (Adjective)	Watchful, cautious
130.	Civil (Adjective)	Polite
131.	Coherent (Adjective)	Holding together, make sense
132.	Congregate (Verb)	To come together
133.	Construe (Verb)	To interpret
134.	Arcane (Adjective)	Mysterious
135.	Demur (Verb)	To disagree
136.	Despondent (Adjective)	Extremely unhappy
137.	Atrocious (Adjective)	Extremely unpleasant

Enhance Your Word Power		
Sl.	**Word**	**Meaning**
138.	Decimate (Verb)	To destroy
139.	Militate (Verb)	To work against
140.	Miscreant (Noun)	Criminal
141.	Mundane (Adjective)	Ordinary
142.	Neophyte (Noun)	Novice, beginner
143.	Impunity (Noun)	Freedom from punishment
144.	Inate (Adjective)	Inborn, existing from birth
145.	Ossify (Verb)	To change into bone
146.	Inchoate (Adjective)	Undeveloped
147.	Fish (Verb)	To seek indirectly
148.	Folly (Noun)	Lack of good sense
149.	Wager (Verb)	To risk something
150.	Lapse (Noun)	Accidental mistake, slip, error
151.	Hamper (Verb)	To interface
152.	Insurgent (Noun)	A person who revolts against the government
153.	Integral (Adjective)	Essential
154.	Laconic (Adjective)	Using a few words
155.	Lampoon (Verb)	To make, to satirize
156.	Maudlin (Adjective)	Silly and sentimental
157.	Mendacious (Adjective)	Dishonest
158.	Mendicant (Noun)	Beggar

Enhance Your Word Power		
Sl.	**Word**	**Meaning**
159.	Mortify (Verb)	To humiliate
160.	Novel (Adjective)	New, original
161.	Cryptic (Adjective)	Mysterious
162.	Castigate (Verb)	To criticise
163.	Chagrin (Noun)	Humiliation
164.	Corroborate (Verb)	To confirm
165.	Copious (Adjective)	Abundant, Plentiful
166.	Disparate (Adjective)	Different, Unequal
167.	Efface (Verb)	To erase
168.	Egregious (Adjective)	Extremely bad
169.	Eminent (Adjective)	Well-known
170.	Espouse (Verb)	To support
171.	Forgo (Verb)	To do without
172.	Forsake (Verb)	To abandon, to remove
173.	Furtive (Adjective)	Secretive, sly
174.	Founder (Verb)	To fail, to collapse
175.	Revere (Verb)	To respect highly
176.	Rustic (Adjective)	Rural
177.	Rudimentary (Adjective)	Basic
178.	Sloth (Noun)	Laziness
179.	Taciturn (Adjective)	Untalkative by nature
180.	Tangible (Adjective)	Touchable

Enhance Your Word Power		
Sl.	**Word**	**Meaning**
181.	Tenacious (Adjective)	Stubborn
182.	Temperate (Adjective)	Mild, moderate
183.	Minuscule (Adjective)	Very tiny
184.	Sanctity (Noun)	Purity, sacredness
185.	Heed (Verb)	To pay attention
186.	Corybantic (Adjective)	Wild, uncontrolled
187.	Pinnacle (Noun)	The highest point
188.	Niggle (Verb)	To keep complaining about small things
189.	Appease (Verb)	To pacify
190.	Prenuptial (Adjective)	Before marriage
191.	Lament (Verb)	To express sorrow
192.	Austere (Adjective)	Extremely simple
193.	Dulcet (Adjective)	Sweet
194.	Beget (Verb)	To produce
195.	Cohere (Verb)	To stick together
196.	Roster (Noun)	List
197.	Delve (Verb)	To go deep into
198.	Mutate (Verb)	To change
199.	Commute (Verb)	To travel, to change
200.	Mutual (Adjective)	Reciprocal, common
201.	Moribund (Adjective)	Near death
202.	Mordant (Adjective)	Extremely sharp
203.	Ramble (Verb)	To wonder without a definite route or goal

Enhance Your Word Power		
Sl.	Word	Meaning
204.	Precipitous (Adjective)	Very steep
205.	Recapitulate (Verb)	To summarise the main points
206.	Muster (Verb)	Together
207.	Ensanguined (Adjective)	Covered with blood
208.	Fidelity (Noun)	Faithfulness
209.	Strand (Verb)	To leave in a helpless manner
210.	Amity (Noun)	Friendship
211.	Inimical (Adjective)	Unfriendly, hostile
212.	Profess (Verb)	To claim, to declare openly
213.	Bellicose (Adjective)	Eager to fight
214.	Sophisticated (Adjective)	Very complex or complicated
215.	Fancy (Verb)	To imagine
216.	Queer (Adjective)	Strange
217.	Blanch (Verb)	To become white
218.	Peroration (Noun)	The conclusion of a speech
219.	Melancholy (Adjective)	Very sad, gloomy
220.	Onslaught (Noun)	A very angry and forceful attack
221.	Scrutinise (Verb)	To examine vary carefully
222.	Serendipity (Noun)	Accidental good fortune
223.	Singular (Adjective)	Unique

Enhance Your Word Power		
Sl.	Word	Meaning
224.	Premise (Noun)	An assumption
225.	Hapless (Adjective)	Unlucky
226.	Hegemony (Noun)	Leadership, especially of nation over another
227.	Chastise (Verb)	To punish
228.	Craven (Adjective)	Cowardly
229.	Blithe (Adjective)	Carefree, Light hearted
230.	Sanguine (Adjective)	Cheerful and optimistic
231.	Abscond (Verb)	To run away from the law
232.	Bilk (Verb)	To deceive, cheat
233.	Purloin (Verb)	To steal
234.	Reprehensible (Adjective)	Deserving rebuke or censure
235.	Tribunal (Noun)	A court of Justice
236.	Venal (Adjective)	Unprincipled, easily bribed
237.	Abysmal (Adjective)	Very profound
238.	Florid (Adjective)	Flowery, reddish
239.	Precarious (Adjective)	Risky
240.	Bravado (Noun)	False courage
241.	Regale (Verb)	To delight, to entertain
242.	Repose (Noun)	Calmness, peace
243.	Slake (Verb)	To satisfy
244.	Pungent (Adjective)	Having a strong, sharp taste or smell
245.	Lucrative (Adjective)	Profitable, beneficial

Enhance Your Word Power		
Sl.	Word	Meaning
246.	Primeval (Adjective)	Ancient,
247.	Abstain (Verb)	To chose to refrain from or to keep away
248.	Apprise (Verb)	To inform
249.	Dissuade (Verb)	To discourage from something
250.	Dither (Verb)	To hesitate
251.	Divulge (Verb)	To disclose, to make something known
252.	Stymie (Verb)	To hinder, to obstruct
253.	Rescind (Verb)	To repeal, to cancel
254.	Extenuate (Verb)	To reduce, the strength, to lessen
255.	Audacious (Adjective)	Bold, daring
256.	Vitriolic (Adjective)	Savagely hostile, bitter
257.	Nonchalant (Adjective)	Indifferent or cool
258.	Morose (Adjective)	Gloomy, sad, sullen
259.	Proscribe (Verb)	To prohibit, forbid
260.	Nullify (Verb)	To make invalid
261.	Umbrage (Noun)	Offence, resentment
262.	Asperity (Noun)	Harshness, roughness of manner
263.	Guffaw (Noun)	A noisy, coarse burst of laughter
264.	Rigmarole (Noun)	Rambling or confusing talk
265.	Rubric (Noun)	A class or category
266.	Sundry (Adjective)	Various, Miscellaneous

Enhance Your Word Power		
Sl.	Word	Meaning
267.	Stringent (Adjective)	Very strict, Harsh
268.	Plaintive (Adjective)	Mournful
269.	Expedient (Adjective)	Appropriate for a purpose
270.	Abhor (Verb)	To hate, detest
271.	Aficionado (Noun)	A fan or devotee
272.	Altercation (Noun)	A heated dispute or quarrel
273.	Placid (Adjective)	Calm and peaceful
274.	Rancor (Noun)	A bitter feeling of ill will.
275.	Pillage (Verb)	To rob of goods in the time of war.
276.	Contentious (Adjective)	Quarrelsome
277.	Fervor (Noun)	Zeal, intense emotion
278.	Debacle (Noun)	A sudden disaster
279.	Frugal (Adjective)	Careful and economical
280.	Opulent (Adjective)	Affluent, possessing great wealth
281.	Profligate (Adjective)	Wasteful, extravagant
282.	Obviate (Verb)	To get rid of
283.	Elite (Noun)	The best or most skilled members of a group
284.	Apex (Noun)	The highest point
285.	Embargo (Noun)	An official ban on trade
286.	Fetid (Adjective)	Having a bad smell

Enhance Your Word Power		
Sl.	**Word**	**Meaning**
287.	Fetter (Verb)	To bind, to confine
288.	Fidelity (Noun)	Loyalty
289.	Flout (Verb)	To treat scornfully
290.	Aspersion(Noun)	An act of defamation
291.	Fusion (Adjective)	Union, synthesis
292.	Halcyon (Adjective)	Calm and peaceful, happy
293.	Impinge (Verb)	To strike, encroach
294.	Incarnate (Adjective)	Having bodily form
295.	Lascivious (Adjective)	Lustful
296.	Latent (Adjective)	Hidden
297.	Laud (Verb)	To praise
298.	Libido (Noun)	Sexual desire
299.	Lilliputian (Adjective)	Extremely small
300.	Limpid (Adjective)	Clear, transparent
301.	Nugatory (Adjective)	Invalid
302.	Cardinal (Adjective)	Of foremost importance
303.	Accrue (Verb)	To accumulate
304.	Artifact (Noun)	Item made by human craft
305.	Argot (Noun)	A specialized vocabulary used by a group
306.	Avarice (Noun)	Greed
307.	Bedizen (Verb)	To dress in a vulgar, showy manner

Enhance Your Word Power		
Sl.	**Word**	**Meaning**
308.	Behemoth (Noun)	Huge creature
309.	Bifurcate (Verb)	To divide into two parts
310.	Blandishment (Noun)	Flattery
311.	Bolster (Verb)	To support
312.	Boorish (Adjective)	Rude, insensitive
313.	Burnish (Verb)	To polish
314.	Cantankerous (Adjective)	Irritable, ill humored
315.	Champion (Verb)	To defend or support
316.	Centrifugal (Adjective)	Moving away from the centre
317.	Centripetal (Adjective)	Moving towards the centre
318.	Compendium (Noun)	Comprehensive summary
319.	Concoct (Verb)	To invent
320.	Coquette (Noun)	A woman who flirts
321.	Erudite (Adjective)	Learned, expert, scholarly
322.	Ignominious (Adjective)	Shameful, disgraceful
323.	Essay (Verb)	To make an attempt
324.	Eulogy (Noun)	High praise
325.	Euthanasia (Noun)	Mercy killing
326.	Expatiate (Verb)	To speak or write at length
327.	Expatriate (Verb)	To send into exile
328.	Explicate (Verb)	To explain, interpret

Enhance Your Word Power		
Sl.	Word	Meaning
329.	Expository (Adjective)	Explanatory
330.	Extirpate (Noun)	To destroy
331.	Fauna (Noun)	Animal of a period or region
332.	Gainsay (Verb)	To deny, oppose
333.	Incorporate (Verb)	To introduce something into another thing already in existence.
334.	Labile (Adjective)	Likely to change
335.	Occlude (Verb)	To shut, block
336.	Oscillate (Verb)	To move back and forth
337.	Ostentatious (Adjective)	Showy
338.	Paean (Noun)	Song of joy
339.	Rail (Verb)	To scold with abusive language
340.	Sedulous (Adjective)	Diligent
341.	Spendthrift (Noun)	A person who spends money recklessly
342.	Squalor (Noun)	Filthy, dirty, sordid
343.	Tantamount (Adjective)	Equal to, similar to
344.	Vista (Noun)	A view or prospect
345.	Rancid (Adjective)	Rotten
346.	Desiccate (Verb)	To dry out thoroughly
347.	Vapid (Adjective)	Dull, tedious
348.	Impel (Verb)	To motivate
349.	Quibble (Verb)	To find fault

Enhance Your Word Power		
Sl.	Word	Meaning
350.	Veracious (Adjective)	Honest, truthful
351.	Nepotism (Noun)	Favoritism for kin when conferring jobs
352.	Tentative (Adjective)	Uncertain, not fixed
353.	Nomadic (Adjective)	Wadering, roaming from place to place
354.	Obsolete (Adjective)	Out of use
355.	Obliterate (Verb)	To erase, destroy
356.	Lackluster (Adjective)	Dull, lacking liveliness
357.	Transgress (Verb)	Beyond the limit
358.	Coddle (Verb)	To cook in water just below the boiling point
359.	Expound (Verb)	To explain in detail
360.	Macabre (Adjective)	Gruesome horrible
361.	Pertinent (Adjective)	Applicable
362.	Raddle (Verb)	To twist together
363.	Witless (Adjective)	Foolish, or silly
364.	Pavid (Adjective)	Timid, fearful
365.	Pristine (Adjective)	Pure, fresh, clean
366.	Vindictive (Adjective)	Revengeful, hateful
367.	Reluctant (Adjective)	Unwilling to do something
368.	Plagiarise (Verb)	To steal words, thoughts in literary composition

Enhance Your Word Power		
Sl.	Word	Meaning
369.	Jeopardize (Verb)	To put at risk
370.	Prevalent (Adjective)	Widespread, widely accepted
371.	Accost (Verb)	To approach and speak to someone, welcome
372.	Unison (Noun)	A condition of perfect agreement.
373.	Deride (Verb)	To treat with contempt
374.	Encomium (Noun)	A formal expression of praise
375.	Vogue (Noun)	The prevalent way or fashion
376.	Overt (Adjective)	Apparent, clear, obvious
377.	Posh (Adjective)	Elegant and fashionable
378.	Discomfit (Verb)	To make uneasy
379.	Ardent (Adjective)	Intense, fervent,
380.	Satiate (Verb)	To Satisfy
381.	Knell (Noun)	The sound of a bell
382.	Ratify (Verb)	To make valid
383.	Imbue (Verb)	To inspire, Pervade
384.	Votive (Adjective)	Dedicated by a vow

Enhance Your Word Power		
Sl.	Word	Meaning
385.	Nefarious (Adjective)	Wicked, evil
386.	Judicious (Adjective)	Wise, Prudent
387.	Procrastinate (verb)	To delay
388.	Probe (Verb)	To examine thoroughly
389.	Proficient (Adjective)	Adept, well versed
390.	Rant (Verb)	To speak loudly
391.	Rapport (Noun)	Good relationship
392.	Arable (Adjective)	Suitable for cultivation
393.	Gelid (Adjective)	Extremely cold
394.	Extricate (Verb)	To release from difficulty
395.	Nonplussed (Adjective)	Confused, perplexed
396.	Transpire (Verb)	To come to pass
397.	Jostle (Verb)	To push
398.	Peerless (Adjective)	Matchless, unique
399.	Stern (Adjective)	Harsh, rigorous
400.	Implore (Verb)	To request
401.	Indigence (Noun)	Poverty

CHAPTER 2

Idioms and Phrases

Sl.	Idioms	Meaning
1.	Under a cloud	Under suspicion
2.	Flogging a dead horse now	Wasting time in useless effort
3.	Look down upon	Regard with contempt
4.	Face the music	Get reprimanded
5.	Token strike	Short strike held as a warning
6.	Fell foul of	Felt annoyed with
7.	By fits and starts	Irregularly
8.	Enough rope	Enough freedom for action
9.	At sea	Puzzled
10.	Sweeping statement	Generalised statement
11.	A big cheese	Important person
12.	Go back upon or from	Fail to keep
13.	To drive to the wall	To crush
14.	Dutch courage	Courage left by strong drink
15.	Pine away	Fall ill and die on account of sorrow
16.	Be quite of somebody	Be free from
17.	Show the white feather	Run away
18.	Look daggers	Look angrily
19.	Rub shoulders	Mix
20.	Beauty sleep	Sleep before midnight
21.	Like and wind	Very fast
22.	Play ducks and drakes	Destroy
23.	Go to roost	Go to bed
24.	Black ox	Bad luck
25.	Rave up	Exciting party
26.	Gallop ahead of somebody	Improve fast
27.	Oodles of something	Much
28.	Laugh in one's sleeve	Laugh secretly
29.	Sackcloth and ashes	Lamentation
30.	Kangaroo court	Non-government court
31.	Haul over the coals	Chide
32.	Mother wit	General sense
33.	Elbow room	Supply with enough space
34.	An old flame	Lower
35.	Throw mud at	Provoke
36.	Inkling of something	A little knowledge something
37.	Bury the hatchet	End quarrel
38.	Tom, Dick and Harry	Ordinary

Sl.	Idioms	Meaning
39.	Shot in the locker	No money in one's pocket
40.	Sharp practice	Cheating, perception
41.	Point blank	Directly
42.	Itching palm	Habits of taking bribe
43.	On the cards	Certain
44.	Miss the boat	It is too late
45.	Given up	Abandoned
46.	Done to death	Murdered
47.	Rub him the wrong way	Annoy him
48.	Head and Front	The main part
49.	In black and white	In writing
50.	Fair and Square	Honest
51.	By long odds	Without fail
52.	Black comedy	Comedy with a tragic element/Basic pessimism
53.	At a pinch	In case of emergency
54.	At a tender age	Still very young
55.	Bosom friend	Close friend
56.	Cast off	Reject
57.	Do up	Arrange
58.	Cave in	Fall
59.	Have a go	Try
60.	A going concern	Business in working order
61.	Go it alone	Act by oneself
62.	Gala day	Happy day
63.	Grapes from thorns	Good from bad
64.	In the air	Not decided
65.	On the job	in action
66.	Jar upon	Irritate
67.	Knock out	Defeat
68.	Raised the banner	Lead
69.	Rundown	Kill
70.	On the ball	Doing a good job
71.	Rabbit hate	Great hate
72.	Pay one's way	Not get into debt
73.	Run off at the mouth	To talk too much
74.	Down and out	Homeless, A person without any means or support
75.	At dead of night	Midnight
76.	Dead from the neck up	Extremely stupid
77.	Stone deaf	Completely deaf
78.	Dead to the world	In a very deep sleep
79.	Cut a dash	To wear very good clothes/dress to kill
80.	The very devil	Very difficult
81.	Easy on the eye	Very attractive
82.	Eternal rest	death
83.	Fair and square	Honestly
84.	My hands are full	Very busy
85.	Pennies from heaven	Sudden money
86.	Pop the questions	To ask someone to marry
87.	Have no truck with someone	To have no contact

Sl.	Idioms	Meaning
88.	Draw the veil over	Not to discuss anything
89.	Carry the can	To take the blame
90.	At eyes	To watch something attentively
91.	All in	Exhausted
92.	Down hearted	Sad, gloomy
93.	Head for something	To move towards a place
94.	Pie in the sky	An unrealistic hope
95.	Pay lip service	To agree by word of mouth only
96.	The fact of the matter	Simple truth
97.	Fair sex	Ladies
98.	Without fail	Certainly
99.	Play fast and loose	To be unreliable
100.	Feel at home	Feel easy
101.	Go ahead	Make progress
102.	Grass on somebody	Inform the police
103.	Green fingers	Skilled gardening
104.	Grow in to something	Become with the passage of time
105.	Eat out of somebody's hand	Have complaint towards somebody
106.	Keen eye	Fully alive
107.	Keep something from somebody	Not tell somebody
108.	Look over	Examine
109.	Make over	Deliver
110.	Beyond measure	Limit less

Sl.	Idioms	Meaning
111.	People of quality	Upper class
112.	In queer street	In debt
113.	Set in	Begin
114.	On the round	Starting a new relation
115.	To can one's salt	To be one's guest
116.	Vest with	Furnish
117.	Come to view	To be seen
118.	Point of view	Opinion
119.	Wade into	Attack
120.	Look blue	Upset
121.	Shed crocodile tears	False tears of sorrow
122.	The tip of my tongue	About to say
123.	Going against the grain	Doing things differently from what you usually do
124.	Art of designing or writing on way	Cerography
125.	Word of mouth	Orally
126.	Taken to task	Scolded
127.	Water under the bridge	Something I can't change
128.	Stick to my guns	Hold on to my decision
129.	About time	To finally do something
130.	Ball park figure	An approximate amount
131.	Bear fruit	Lead to the result you expected
132.	Black hole	A large gap

Sl.	Idioms	Meaning
133.	A hard nut to crack	A prob that is difficult to solve
134.	Cash cow	A product or service that makes a lot of money
135.	Crunch time	On important time when a decision has to be made
136.	Dirt cheap	Extremely cheap
137.	Drive a hard bargain	Negotiate tough conditions
138.	Feather your nest	Making sure you have or make enough money
139.	Gather pace	Speed up
140.	Get the show on the road	Get started
141.	Have a quick word	Take to someone briefly
142.	Big cheese	An important person
143.	Bite the butter over with	To get something
144.	Dig your heals in	Refuse to do something or be persuaded
145.	Fat cats	Top people in a company who earn a lot of money
146.	Get into the water	Get into trouble
147.	Donkey work	Difficult or less glamorous parts of the job
148.	Drag your feet	Making a decision
149.	Corporate clone	A company man
150.	Go belly up	Fail
151.	Golden handshake	A large amount of money given to someone when they leave the job

Sl.	Idioms	Meaning
152.	Hit a dead lock	Unable to progress further
153.	Jump the gun	To start something too early
154.	Keep something afloat	Keep something going
155.	Fall on hard times	Be experiencing difficult times
156.	Dish the dirt	Reveal all the secretes
157.	Cut back	Reduce spending
158.	Break the bank	Be too expensive to buy
159.	Roll over	Transfer from one bank/company
160.	Rush off your feet	Be very busy
161.	Saddled with debt	Have too much debt to pay
162.	Sweeten the deal (a sweetener)	An incentive to agree a deal
163.	Take home (take home pay)	The amount of money are paid
164.	Get the wrong end of the stick	Misunderstand something
165.	Face the music	To face a problem and not deny that it exists
166.	Gain ground	Make progress
167.	Put the brakes on	Slow something down
168.	Pin money	Small amount of money (originally money given by a husband to his wife for her personal use)
169.	Play the game	Accept the rules
170.	On-off	When something stops and stars

Sl.	Idioms	Meaning
171.	Run on empty	With no new ideas, energy or financing
172.	See off	Succeed against competition to win the bid for the project.
173.	Wriggle room	Room for maneuver
174.	Write down	Reduce/agree that the tax
175.	Wind up	Close a company
176.	Horse of a different color	Guide a different matter
177.	Let the cat out of the bag	Inform before hand
178.	For the birds	Uninteresting and meaningless
179.	Straight from, the horse's mouth	From a reliable source
180.	Horse around	Play around
181.	Cat got your tongue	Can't talk
182.	Get in someone's hair	Bother someone
183.	Shoot off one's mouth	Express one's opinions loudly
184.	Jump down someone's throat	Become angry with someone
185.	Pay through the nose	Pay too high, a price
186.	Bumpy ride	Experiencing a lot of problems
187.	Churn rate	Rate at which you win and lose customers
188.	Hard nosed	Business minded
189.	Latched on to	Promoted
190.	To fight shy of	To avoid
191.	To go to the dogs	To be ruined

Sl.	Idioms	Meaning
192.	Rose coloured glasses	A positive outlook on life
193.	A ripe old age	Very old age
194.	Back to the grind stone	Back to work
195.	Have a ball	To have a very enjoyable time
196.	To foam at one's month	To get very angry
197.	Achilles' heal	Weak point
198.	To raise adjust	To create confusion
199.	Come cap in hand	Very humble
200.	To play ducks and drakes	To squander money
201.	Man of letters	Scholar
202.	At stako	In danger
203.	To beat about the bush	To evade the topic
204.	Cut the gordian knot	To perform a difficult task
205.	Through thick and thin	Under all circumstances
206.	To kick up a row	To make great noise
207.	A lick and a promise	To do a job in a careless manner
208.	Be all ears	To listen to carefully
209.	American dream	To lead a life of success
210.	Arms of morpus	Sound sleep
211.	Band aid solution	A temporary solution
212.	Bring home the bacon	To provide food for the family
213.	Chip off the old block	A son like or be behaves like his own father

Sl.	Idioms	Meaning
214.	Dress up to the nines	Wearing best clothes
215.	To get up on the wrong side of the bed	To be in a bad mood for no obvious reason
216.	Half beated	Courageless
217.	Eat out of some body's hand	Have complaint towards somebody
218.	Hair breadth escape	Narrow escape
219.	Have away	Remove
220.	The grass roots	Basic level
221.	Lay a ghost	Exercise an evil spirit
222.	The gift on the gab	Ability to speak fluently
223.	Good money	Genuine
224.	Guard against	Defence
225.	Jot walk	Slow but regular walk
226.	In vogue	Popular
227.	Lay somebody or something	Bet
228.	Over and over	In addition to
229.	On the whole	Wholly
230.	By fair or foul means	Any how
231.	In meal or malt	In one way or the other
232.	Open up	Start a business
233.	Mend or end	Improve or finish
234.	Spin a yarn	Tell tales
235.	Way a war	Launch a war
236.	Work on	Continue
237.	As sure as gun	Quite sure

Sl.	Idioms	Meaning
238.	In a tight corner	In a difficult situation
239.	Wild goose chase	Futile search
240.	Does not hold water	Can't be believed
241.	Out and out a lie	Totally
242.	A while elephant	Costly or trouble some possession
243.	Fair and square	Honest
244.	No love lost	Not on good terms
245.	All at once	Suddenly
246.	All a discount	Poorly esteemed
247.	By fits and starts	Impassively and irregularly
248.	In great difficulties	In deep water
249.	Ins and outs	Full details
250.	Out of the blue	Unexpectedly
251.	Cry in the wilderness	In effective demand
252.	Herculean task	A difficult task
253.	Fight with their back to the wall	Fight desperately
254.	Under the sun	Any where
255.	End up somebody with something	Be of good quality
256.	Lumber of with	Full of
257.	In retrospect	Recall the past event
258.	To be at the zenith of	Be at the peak of
259.	Get out of hand	Get out of control
260.	Hit the scale	To go to sleep

Sl.	Idioms	Meaning
261.	Keep a good table	Give luxurious meal
262.	Sow one's wild oats	Involve in lust
263.	Eke out	Make up
264.	At random	Without purpose
265.	At stake	In danger
266.	At the eleventh hour	At the last moment
267.	At hand	Neat
268.	At a tender age	Still very young
269.	Aghast	Filled with fear
270.	Bear a hand	Help
271.	Beat off	Face boldy
272.	Sowing wild oats	Irresponsible pleasure seeking in young
273.	Axe to grind	A private interest to serve
274.	Reading between the lines	Looking for meanings that are not actual
275.	To give the devil his due	To give credit to even a notorious person
276.	To eat humble pie	To have to apologise
277.	Hobson's choice	No real choice at all
278.	With a grain of salt	To listen to something with considerable doubt
279.	Tax one's patience	Very hard test of patience
280.	Save one's skin	Save oneself
281.	Carry the day	To win
282.	Cast away	Throw oft, to give up
283.	Call on	Visit to person
284.	Draw hear	Come near
285.	Due	To be paid
286.	Feel one's ears burning	Feel one is being talked about
287.	Ear-short	Hearing distance
288.	Lend a ear	to listen
289.	Take the edge off	Make blunt
290.	Elixir of life	A drag for prolonging life
291.	Duck diamond	Beloved
292.	Every now and then	Often
293.	Fall on easy prey	Become victim easily
294.	Fall foul	To make an attack
295.	Fairy tall	An incredible tale
296.	Far and wide	To far places
297.	Fight tooth and nail	To fight very fiercely
298.	Have a bone to pick with	An unpleasant issue
299.	Goose egg	Zero
300.	A wild goose chase	A useless attempt

Practice Exercise Idioms

Set-1 IDIOMS

1. Go to the dogs
(a) Go to ruin (b) Go to heaven
(c) Be ashamed of (d) to resign

2. Come off with flying colours
(a) Compensate
(b) Take care of
(c) Achieve creditable success
(d) Keep silent

3. Hit below the belt
(a) To act unfairly
(b) To act in fair manner
(c) To give up
(d) To act bravely

4. Cut a sorry figure
(a) To beat severely (b) Make a poor show
(c) To venture (d) To discourage

5. Leave no stone unturned
(a) To rest
(b) To do everything that can be done
(c) To fight
(d) To punish

6. Make a clean breast of
(a) To blame others
(b) To discourage
(c) To respect
(d) To confess fully one's fault

7. Over head and ears
(a) Very tall
(b) Unconcerned
(c) Completely immersed
(d) To quarrel

8. Red-letter day
(a) A day of great joy or importance
(b) A miserable day
(c) Ripe time
(d) In perfect order

9. Make hay while the sun shines
(a) To dance happily
(b) To hurry
(c) Take advantage of favourable condition
(d) To destroy

10. To let the cat out of the bag
(a) To get into trouble (b) To keep a secret
(c) Have a fever (d) To reveal a secret

11. Laugh up one's sleeve
(a) To amuse
(b) To make other laugh
(c) To laugh secretly
(d) To laugh in pubic

12. Take the bull by the horns
(a) To evade a difficulty
(b) Blush or grow pale
(c) Grapple with difficulty
(d) Spoil with flattery

13. Throw out of gear
(a) To replace (b) Disturb
(c) To decide (d) Take up tune

14. Turn over a new leaf
(a) To mend one's way (b) To be present
(c) To discard (d) To survive

15. Burn one's fingers
(a) Become reconciled (b) To hurry
(c) To get into trouble (d) Make it difficult for

16. To hang in the balance
(a) To guess right (b) To manage to live
(c) To be undecided (d) To withdraw

17. To leave in the lurch
(a) To study
(b) To leave in difficulties
(c) To lay aside
(d) To face the difficulty

18. To pay one back in one's own coin
(a) To resemble (b) Behave naturally
(c) Discover a way (d) To return tit for tat

19. To get away
(a) To escape (b) To face
(c) To pass (d) To reach

20. Cheek by jowl
(a) Fighting (b) screaming
(c) Close together (d) Rubbing cheek

21. To given the cold shoulder
(a) To be unfriendly
(b) To abandon someone
(c) To abuse someone
(d) To curse someone

22. Touch act to follow
(a) Acting like a tough man
(b) Difficult to perform like that
(c) Difficult to act well
(d) Acting skill is tough

23. Took to her heels
(a) bent down
(b) run away
(c) kicked off her shoes
(d) rubbed her heels

24. Turn up
(a) twist (b) curt
(c) appear (d) wind

25. Bad blood
(a) Low blood pressure
(b) Feeling of love
(c) Feeling of hatred
(d) High blood pressure

Set-2 IDIOMS

1. I have recently changed my job and I am doing to have teething problems
(a) difficulties at the start
(b) difficulties at the end
(c) difficulties all the time
(d) problem with my teeth

2. The soldiers fought tooth and nail to save their country.
(a) using unfair means
(b) with strength and fury
(c) with weapons
(d) as best as they could

3. To bury the hatchet
(a) To obtain money
(b) To make friends
(c) To keep a secret
(d) To make peace

4. Watch his step
(a) Behave with caution
(b) Observe the threshold
(c) Walk carefully
(d) Watch his friends

5. To beat about the bush
(a) To cut down trees
(b) To plants more trees
(c) To hurt someone
(d) Not to come to the point

6. Unless you grease his palms he will not do your work
(a) talk to him (b) flatter him
(c) beat him (d) bribe him

7. The police closed the book on the murder case
(a) Solved the case of
(b) Stopped working on
(c) Handed the case over to another agency
(d) Refused to take up

8. His arguments cut no ice with me
(a) had no influence on me
(b) did not hurt me
(c) did not benefit me
(d) did not make me proud

9. To hit the jackpot
(a) To gamble
(b) To make money unexpectedly
(c) To inherit money
(d) To become bankrupt

10. To throw dust in one's eyes
(a) To show false things
(b) To make blind
(c) To harm someone
(d) To deceive

11. A stop to cerberus
(a) Bribery
(b) I lush money
(c) Ransom to an enemy
(d) Money for compensation

12. To go at each other hammer and tongs
(a) To fight with weapons
(b) Doing with great enthusiasm or energy
(c) To hug each other
(d) To go crazy about meeting each other

13. A chip off the old block
(a) A piece of wood
(b) An old friend
(c) Characteristics of one's ancestors
(d) A good bar gain

14. To give / get the bird
(a) To get the awaited
(b) To have good luck
(c) To shout
(d) To get the impossible

15. Pell - mell
(a) I hurried disorder
(b) Gaiety and excitement
(c) Extremely worried
(d) None of these

16. Tall order
(a) Simple
(b) Customary
(c) Easy
(d) Too much difficult to do or get

17. Was plain sailing
(a) was very easy
(b) was complicated
(c) was competitive
(d) was uncomfortable

18. A bolt from the blue
(a) Struck by thunder
(b) A piece of bad luck
(c) A flash of lighting
(d) A complete surprise

19. To carry off the bell
(a) To call other for help
(b) To beg the first position
(c) To steal all wealth and flee
(d) To trouble others

20. To see eye to eye with
(a) To agree (b) To stare fixedly
(c) To be angry (d) To take revenge

21. To take a leap in the dark
(a) To take risk
(b) To hazard one self
(c) To do a task secretly
(d) To do a hazardous thing without any idea of the result

22. A man of straw
(a) A man of no substance
(b) A very active person
(c) A worthy fellow
(d) An unreasonable person

23. To push somebody to the wall
(a) To defeat him (b) To humiliate him
(c) To hackle him (d) To knock him down

24. To get into hot waters
(a) To be impatient
(b) To suffer a huge financial loss
(c) To get into trouble
(d) To be in a confused state of mind

25. To play second fiddle
(a) To be happy, cheerful and healthy
(b) To reduce important of one's senior
(c) To support the role and view of another person
(d) To do back seat driving

Answer Key **Idioms : SET-1**

1. (a)	2. (c)	3. (a)	4. (b)	5. (b)	6. (d)
7. (c)	8. (a)	9. (c)	10. (d)	11. (c)	12. (c)
13. (b)	14. (a)	15. (c)	16. (c)	17. (b)	18. (d)
19. (a)	20. (c)	21. (a)	22. (b)	23. (b)	24. (c)
25. (c)					

Answer Key **Idioms : SET-2**

1. (a)	2. (b)	3. (d)	4. (a)	5. (d)	6. (d)
7. (b)	8. (a)	9. (b)	10. (d)	11. (a)	12. (b)
13. (c)	14. (c)	15. (a)	16. (d)	17. (a)	18. (d)
19. (a)	20. (a)	21. (d)	22. (d)	23. (a)	24. (c)
25. (c)					

CHAPTER 3

Words Often Confused

Words Often Confused

1. Accede/Exceed

Accede (V) ➤ *To agree*

Use : She acceded to the demands of the students.

Exceed (V) ➤ *To go beyond the limit.*

Use : Please do not exceed the speed limit.

2. Adverse/Averse

Adverse (Adj) ➤ *Unfavourable*

Use : Most birds can be seen flaying in adverse weather.

Averse (Adj) ➤ *To have a strong feelings against.*

Use : Pecola is averse to smoking with her own husband.

3. Allude/Elude

Allude (Verb) ➤ *To refer indirectly*

Use : Shakespeare did not allude to his poverty in his plays.

Elude (Verb) ➤ *To escape, to avoid*

Use : Hester tried to elude captivity.

4. Alone/Lonely

Alone (Adj) ➤ *Without other people*

Use : I was alone when she entered my office.

Lonely (Adj) ➤ *You are unhappy because you are not with other people.*

Use : I still need you because even today I feed lonely without you.

5. Appraise / Apprise

Appraise (Verb) ➤ *To determine the worth of something.*

Use : I want you to appraise this old car.

Apprise (Verb) ➤ *To inform*

Use : *She was apprised of the death of her mother.*

6. Borrow/Lend

Borrow (Verb) ➤ *To take something from someone for a certain period of time.*
Use : I borrowed forty thousand rupees from my brother.

Lend (V) ➤ *To give something to someone for a certain period of time.*
Use : I can lend you Rs. 5000 on 16 December.

7. Cite/Site

Cite (Verb) ➤ *To quote*
Use : He cited the constitution in his defence.

Site (Noun) ➤ *Location of a town, building etc.*
Use : This site is for our swimming pool.

8. Alternate/Alternative

Alternate (Adj) ➤ *Every other, or occurring by turns.*
Use: Candida comes to meet me on alternate days.

Alternate (Verb) ➤ *To happen or exist one after the other repeatedly.*
Use: She alternated between cheerfulness and deep despair.

Alternative (Noun) ➤ *Choice/Option*
Use: There must an alternative to people sleeping on the streets.

9. Custom/Habit

Custom (Noun) ➤ *A belief that has been established for a long time.*
Use: What customs do you have for new year in your country?

Habit (Noun) ➤ *A regular practice or Tendency*
Use: Sandra is unable to give up the habit of smoking.

10. Freedom/Liberty

Freedom (Noun) ➤ To have power to act or change without constraint.
Use: Today we have the freedom to decide our own future.

Liberty (Noun) ➤ Freedom from bondage
Use: The slaves were set at liberty.

11. Attack/Assault

Attack (Verb) ➤ *To try to hurt or defeat using violence.*
Use : Most wild animals do not attack unless they are provoked.
Note: Attack also means, to criticise.

Assault (Noun) ➤ *A violent physical or verbal attack.*
Use: Racially motivated assaults on Asians are increasing by leaps and bounds.

12. Regard/Regardless

Regard (Verb) ➤ *To have an opinion about someone or something.*
Use: I regard Kamladas as the greatest poet in Indian literature.

Regardless (Adverb) ➤ Not being affected by something.
Use: I kept working hard regardless of my failures.

13. Set/Sit

Set (Verb) ➤ *To put something in a certain place.*
Use: Can you set the table for dinner.

Sit (Verb) ➤ *To be seated*
Use: Please do sit down at that desk.

14. Moan/Mourn

Moan (Verb) ➤ *To make a complaint in an unhappy voice.*
Use: She always moans about the weather.

Mourn (Verb) ➤ To express sorrow because of someone's death.
Use: A number of people mourned the death of John Keats.

15. Wander/Wonder

Wander (Verb) ➤ *To travel aimlessly*
Use : She used to wander through the woods.

Wonder (Verb) ➤ *To feel curious*
Use : I wondered to see her melting in his arms.

16. Good/Well

Good (Adj) ➤ *Pleasant or enjoyable. Good usually defines the behaviour of a person.*
Use : He is good, so you can count on him.

Well (Adj) ➤ *In good health*
Use : I am well.

17. Bare/Bear

Bare (Adjective) ➤ *Without covering or clothing,*
Use : I saw him go bare foot.

Bear (Noun) ➤ *A big growly wild animal.*
Use : I saw a bear in the zoo yesterday.

18. Brake/Break

Brake (Noun) ➤ *A device which is used to slow or stop a vehicle.*
Use : You should apply breaks to slow down.

Break (Verb) ➤ *To smash or divide into parts violently.*
Use : Do not break the glass please.

19. Advice/Advise

Advice (Noun) ➤ *Guidance or Recommendations*
Use : He gave me so many pieces of advice yesterday.

Advise (Verb) ➤ *To give someone advice*
Use : Candida advised me to compose a sonnet.

20. Censor/Sensor

Censor (Verb) ➤ *To examine (a book, film etc.) Officially and suppress unacceptable parts of it.*

Use : The book was heavily censored when first published.

Sensor (Noun) ➤ *A device that is used to record that something is present.*

Use : The security device has a heat sensor which detects the presence of people and animals.

21. Affect/Effect

Affect (Verb) ➤ *To influence*

Use : Smoking may affect your health adversely.

Effect (Noun) ➤ *Result*

Use : What are the effects of war?

22. Fair/Fare

Fair (Adj) ➤ *Beautiful*

Use : You look fair in black saree.

Fare (Noun) ➤ *Charges*

Use : What is the fare for Delhi?

23. Seem/Seam

Seem (Verb) ➤ *Appear*

Use : She seems to be a stunning woman.

Seam (Noun) ➤ *Split*

Use : There was a seam in the Himalayas.

24. Pray/Prey

Pray (Verb) ➤ *Address a prayer to God.*

Use : I pray of God everyday.

Prey (Verb) ➤ *To hunt and kill for food.*

Use : The survey claims loan companies prey on weak families already in debt.

25. Die/Dye

Die (Verb) ➤ *To pass away*

Use : She died in London ten years ago.

Dye (Verb) ➤ *To colour*

Use : I dyed my hair black.

26. Accuse/Allege

Accuse (Verb) ➤ *To blame, or to bring a charge against*

Use : He was accused of murder.

Allege (Verb) ➤ *To claim something that has not been proven.*

Use : It was alleged that he secretly cooperated with the English during the war.

27. Healthful/Healthy

Healthful (Adj) ➤ *Conducive to good health*
Use : Doctor advises us to eat healthful food.
Note : Remember: Things are healthful and people are healthy.

Healthy (Adj) ➤ *In good health*
Use : A good diet keeps you healthy.

28. Learn/Teach

Learn (Verb) ➤ *To acquire knowledge*
Use : I want to learn how to speak Russian.

Teach (Verb) ➤ *To impart knowledge*
Use : One of my friends used to teach me mathematics.

29. Persecute/Prosecute

Persecute (Verb) ➤ *To subject to ill treatment.*
Use : He was not understood by the society, so he was persecuted.

Prosecute (Verb) ➤ *To bring, a lawsuit against*
Use : The former president was prosecuted for the anomalies during her administration.

30. Lie and Lay

Lie (Verb) ➤ *To rest. It is an intransitive verb and never has an object.*
Use : I am going to lie down for an hour.

Lay (Verb) ➤ *To put*
Use : Please lay out the books on the table.

31. Hanged/Hung

Hanged (Verb) ➤ *Hanged (Verb) is used when referring to someone being sentenced to death via hanging.*
Use : No one has been hanged in Delhi since 1990.

Hung (Verb) ➤ *Suspend or be suspended from above with the lower part dangling free.*
Use : We hung the picture on the wall.

32. Steal/Rob

Steal (Verb) ➤ *To take something without permission or legal right and without intending to return it.*
Use : Someone has stolen her purse.

Rob (Verb) ➤ *To take money or property from a place or person often using violence.*
Use : They robbed a bank last night.

33. Drown/Sink

Drown (Verb) ➤ *To die in water*
Use : Lydia drowned in the river.

Sink (Verb) ➤ *To go down to the bottom of water.*
Use : The ship sank in the ocean.
Note : Drown refers to living being whereas sink refers to thing.

34. Like/Love

Like (Verb) ➤ *To show that you think something is good.*
Use : I like your haircut.

Love (Verb) ➤ *To feel deep affection for someone.*
Use : I love you head over heels, so I need you for me forever.
Note : Both the verbs can be used for people and things, the only difference is one of degree. Love is much stronger than like .

35. Refuse/Deny

Refuse (Verb) ➤ *Not to take what is offered or not to do what is asked to do.*
Use : Belinda refused to take money.

Deny (Verb) ➤ *To answer in the negative or to say that a statement is not true.*
Use : John denied that he had done it.

36. Wounded/Injured

Wounded (Adj) ➤ *Inflicted with a wound.*
Use : I saw a wounded soldier.

Injured (Adj) ➤ *Harmed, damaged, or physically harmed*
Use : Jack was injured in a car accident.
Note : People are injured as a result of an accident or a fight, but people are wounded in wars and battles.

37. Sick/Ill

Sick (Adj) ➤ *To vomit*
Use : The smell made me sick.

Ill (Adj) ➤ *To be in bad health.*
Use : She has been ill for ten days.

38. Clear/Clean

Clear (Adj) ➤ *Transparent or Unclouded.*
Use : The sky was clear yesterday.

Clean (Adj) ➤ *Not dirty.*
Use : You should keep your hands clean.

39. Customer/Client

Customer (Noun) ➤ *A person who buys goods from shop.*
Use : The grocer has plenty of customers.

Client (V) ➤ *A person or organisation using the services of a lawyer or other professional person or company.*
Use : The lawyer has a number of clients.

40. Finger/Toe

Fingers are on the hand, and toes are on the foot.

☒ **Wrong :** I hurt a finger of my right foot.
☑ **Right :** I hurt a toe of my right foot.

41. Air/Wind

Air / Wind ➤ *Air is what we breathe, and wind is what makes the leaves of the trees move.*

☒ **Wrong** : The strong air blow her hat away.

☑ **Right** : The strong wind blew her hat away.

42. Cost/Price

Cost (Noun) ➤ *The amount paid by the shopkeeper*

Use : How much does it cost?

Price (Noun) ➤ *To amount of money paid by the customer.*

Use : Can you tell me the price of this book?

43. Ground/Floor

Ground (Noun) ➤ *Outside area of the house*

Floor (Noun) ➤ *The part of the room on which we walk*

☒ **Wrong** : When I entered the room, I saw a book on the ground.

☑ **Right** : When I entered the room, I saw a book on the floor.

44. Dress/Suit

Dress (Noun) ➤ *Use for girls' or women's clothes.*

Use : Ms. Dalloway was wearing a green dress.

Suit (Noun) ➤ *For Men's clothes but anyone can wear suits (a Jacket with trousers or a skirt)*

Use : My elder brother bought a new suit last week.

45. Fool/Foolish

Fool (Noun) ➤ *A person who acts unwisely.*

Use : He is a fool.

Foolish (Adj) ➤ *Lacking good sense or judgement, unwise.*

Use : It is foolish to risk skin cancer.

46. Deadly/Dead

Deadly (Adj) ➤ *Causing or able to cause death.*

Use : He was acquitted on changes of assault with a deadly weapon.

Dead (Adj) ➤ *1 No longer alive.* *2. Complete, absolute*

Use : She is dead.

Dead (Adverb) ➤ *Very*

Use : 1. His poems sound dead boring.

2. I am dead sure about it.

47. Climate/Weather

Climate (Noun) ➤ *The weather conditions prevailing in an area over a long period.*

Use : The climate of Delhi is hotter than that of Shimla.

Weather (Noun) ➤ *The conditions in the air above the earth, such as wind, rain or temperature, especially at a particular time over a particular area.*

Use : The weather is expected to remain clear for the next few days.

48. Award/Reward

Award (Noun) ➤ *An award is a prize that you receive for having done something noteworthy.*

Use : The only award that she ever won was the Pulitzer prize.

Reward (Noun) ➤ *A reward is usually associated with something valuable money for example:*

Use : There will be a reward for anyone who will find my missing dog.

49. Graceful/Gracious

Graceful (Adj) ➤ *Behaving in a polite and pleasant way.*

Use : 1. She finally apologized, but she was not very graceful about it.

2. Sheela walked gracefully (adverb) across the room.

Gracious (Adj) ➤ *Kind and pleasant*

Use : She was gracious enough to help me.

50. Moral/Morale

Moral (Noun) ➤ *A lesson that can be derived from a story or experience.*

Use : The moral of the fable was identified by the listeners easily.

Morale (Noun) ➤ *Confidence, enthusiasm*

Use : After losing the game, the team's morale was low.

51. Used to/Be used to

Used to (Modal) ➤ *Is used to indicate past habit.*

Use : Neha used to take tea.

Be used to (Adj) ➤ *Indicates present habit.*

She is used to taking tea.

Note : (be used to + V1 + ing)

be ➤ *is / am / are*

52. Lessen/Lesson

Lessen (Verb) ➤ ***To decrease, mitigate,***

Use : This medicine will lessen your pain.

Lesson (Noun) ➤ *A period of time in which someone is taught.*

Use : I am going to complete this lesson tomorrow.

CHAPTER 4

Synonyms and Antonyms

SYNONYM

A **synonym** is a word or phrase that means exactly or nearly the same as another **lexeme** (word or phrase) in the same language. Words that are synonyms are said to be **synonymous**, and the state of being a synonym is called synonymy. The words begin, start, commence, and initiate are all synonyms of one another.

Words are typically synonymous in one particular sense: for example, long and extended in the **context** long time or extended time are synonymous, but long cannot be used in the phrase extended family.

Synonyms with exactly the same meaning share a **seme** or denotational **sememe**, whereas those with inexactly similar meanings share a broader denotational or **connotational** sememe and thus overlap within a **semantic field**. The former are sometimes called **cognitive synonyms** and the latter, near-synonyms, plesionyms or poecilonyms.

Utilizing synonyms is an elegant and frequently applied method to make a text more pleasant to read. If the same word is used very often in several sentences, it might very well be the case, that this same word should describe something different in each of the sentences. Since synonyms are usually nuanced variations with slightly different meanings, their use will emphasize special characteristics, and thus the text will be more precise.

Tips & Tricks for SOLVE QUESTIONS BASED ON SYNONYM

There's a certain skill involved in choosing the most appropriate synonym; not all are created equal.

It is important to consider the connotation of the word because some synonyms can **inject a different meaning** than what was intended.

Example One synonym for sad is "gloomy." This word carries quite a negative connotation. Depending on the circumstance you can use it but, in this instance, if you just want to say that someone is "down," then another synonym such as "blue" or "unhappy" would be a little less heavy.

- Sort out the words from the options that have a negative impact and those which have a positive effect.
- If we're asked to find the meaning of a positive word, then it would be easier to relate that word with the positive word from the options.
- Now even if we don't know the meaning of any word from the option, we can narrow down the choices to almost half.

List of Synonyms

ACTION —

Come : advance, approach, arrive, near, reach

Go : depart, disappear, fade, move, proceed, recede, travel

Run : dash, escape, elope, flee, hasten, hurry, race, rush, speed, sprint

Hurry : rush, run, speed, race, hasten, urge, accelerate, bustle

Hide : conceal, cover, mask, cloak, camouflage, screen, shroud, veil

Move : plod, go, creep, crawl, inch, poke, drag, toddle, shuffle, trot, dawdle, walk

Do : execute, enact, carry out, finish, conclude, effect, accomplish, achieve, attain

Have : hold, possess, own, contain, acquire, gain, maintain, believe, bear, beget, occupy, absorb, fill, enjoy

Use : employ, utilize, exhaust, spend, expend, consume, exercise

Get : acquire, obtain, secure, procure, gain, fetch, find, score, accumulate, win, earn, rep, catch, net, bag,

Descriptive —

Explain : elaborate, clarify, define, interpret, justify, account for

Idea : thought, concept, conception, notion, understanding, opinion, plan, view, belief

Look : gaze, see, glance, watch, survey, study, seek, search for, peek, peep, glimpse, stare, contemplate, examine, gape

Story : tale, myth, legend, fable, yarn, account, narrative, chronicle, epic, sage, anecdote, record, memoir

Tell : disclose, reveal, show, expose, uncover, relate, narrate, inform

Think : judge, deem, assume, believe, consider, contemplate, reflect, mediate

Feelings —

Anger : enrage, infuriate, arouse, nettle, exasperate, inflame, madden

Angry : mad, furious, enraged, excited, wrathful, indignant, exasperated, aroused, inflamed

Calm : quiet, peaceful, still, tranquil, mild, serene, smooth, composed, collected, unruffled, level-headed, unexcited, detached, aloof

Eager : keen, fervent, enthusiastic, involved, interested, alive to

Fear : fright, dread, terror, alarm, dismay, anxiety, scare, awe, horror, panic, apprehension

Happy : pleased, contented, satisfied, delighted, elated, joyful, cheerful, ecstatic, jubilant, gay, tickled, gratified, glad, blissful, overjoyed

Hate : despise, loathe, detest, abhor, disfavor, dislike, disapprove, abominate

Love : like, admire, esteem, fancy, care for, cherish, adore, treasure, worship, appreciate, savor

Moody : temperamental, changeable, short-tempered, glum, morose, sullen, mopish, irritable, testy, peevish

Sad : miserable, uncomfortable, wretched, heart-broken, unfortunate, poor, downhearted, sorrowful, depressed, dejected, melancholy, glum, gloomy, dismal, discouraged, unhappy

Negative —

Awful : dreadful, terrible, abominable, bad, poor, unpleasant

Bad : evil, immoral, wicked, corrupt, sinful, depraved, rotten, contaminated, spoiled, tainted, harmful, injurious, unfavorable

Dark : shadowy, unlit, murky, gloomy, dim, dusky, shaded, sunless, black, dismal, sad

Dull : boring, tiring, tiresome, uninteresting, slow, dumb, stupid, unimaginative, lifeless, dead

Fat : stout, corpulent, fleshy, beefy, paunchy, plump, full, rotund, tubby, pudgy, chubby, chunky, burly, bulky, elephantine

Gross : improper, rude, coarse, indecent, crude, vulgar, outrageous, extreme, grievous, shameful, uncouth, obscene, low

Hurt : damage, harm, injure, wound, distress, afflict, pain

Lazy : indolent, slothful, idle, inactive, sluggish

Ugly : hideous, frightful, frightening, shocking, horrible, unpleasant, monstrous

Positive —

Brave : courageous, fearless, dauntless, intrepid, plucky, daring, heroic, valorous, audacious, bold, gallant, valiant, doughty, mettlesome

Bright : shining, shiny, gleaming, brilliant, sparkling, shimmering, radiant, vivid, colorful, lustrous, luminous, incandescent, intelligent, knowing, quick-witted, smart, intellectual

Enjoy : appreciate, delight in, be pleased, indulge in, luxuriate in, bask in, relish, devour, savor, like

Funny : humorous, amusing, droll, comic, comical, laughable, silly

Good : excellent, fine, superior, wonderful, marvelous, qualified, suited, suitable, apt, proper, capable, generous, kindly, friendly, gracious, obliging

Great : noteworthy, worthy, distinguished, remarkable, grand, considerable, powerful, much, mighty

Neat : clean, orderly, tidy, trim, dapper, natty, smart, elegant, well-organized, super, desirable, spruce, shipshape, well-kept, shapely

PRACTICE EXERCISE 1

Which of these words is closest in meaning to the word provided?

1. **Captious**
(a) fault finding (b) insecure
(c) changeable (d) vain
2. **Abhor**
(a) argue (b) hate
(c) discover (d) reveal
3. **Graceful**
(a) pretty (b) clever
(c) pleasant (d) present
4. **Trenchant**
(a) incisive (b) greedy
(c) temporary (d) arduous
5. **Cognate**
(a) exult (b) twist
(c) friend (d) relative
6. **Meditative**
(a) oppressed (b) caged
(c) thoughtful (d) happy
7. **Expel**
(a) exile (b) hate
(c) fade (d) clean
8. **Hoax**
(a) malcontent (b) argument
(c) imposter (d) clown
9. **Connubial**
(a) connivance (b) of marriage
(c) encouraging (d) frustrating
10. **Haul**
(a) sleepy (b) crush
(c) proud (d) pull
11. **Jocund**
(a) incredulous (b) merry
(c) revolting (d) dizzy
12. **Pejorative**
(a) arduous (b) disparaging
(c) exorbitantly costly (d) extrovert
13. **Feign**
(a) excite (b) imitate
(c) trick (d) apelike
14. **Magnetism**
(a) ghost (b) force
(c) charm (d) courage
15. **Allocation**
(a) divide (b) decide
(c) cut (d) squabble
16. **Generic**
(a) general (b) cheap
(c) fresh (d) elderly
17. **Qualm**
(a) distress (b) impunity
(c) persevere (d) scruple
18. **Wary**
(a) calm (b) curved
(c) confused (d) cautious
19. **Distort**
(a) wrong (b) evil
(c) deform (d) harm
20. **Sumptuous**
(a) delirious (b) gorgeous
(c) perilous (d) luxurious
21. **Reel**
(a) whirl (b) fish
(c) hit (d) mistake
22. **Inscrutable**
(a) difficult (b) mysterious
(c) inflexible (d) wary
23. **appal**
(a) delirious (b) covered
(c) dismay (d) confuse
24. **Upright**
(a) honourable (b) horizontal
(c) humble (d) supine
25. **Reverie**
(a) palimpsest (b) phantom
(c) daydream (d) curio
26. **Extortionate**
(a) Excessive (b) Kidnap
(c) Extinguish (d) Extract
27. **Loquacious**
(a) talkative (b) thirsty
(c) beautiful (d) complicated
28. **Chimera**
(a) chimney (b) protest
(c) illusion (d) panache
29. **Temerity**
(a) audacity (b) fearfulness
(c) shyness (d) stupidity
30. **Educe**
(a) demand (b) elicit
(c) ideal (d) unlawful
31. **Nabob**
(a) bigwig (b) doubter
(c) frolic (d) converse
32. **Pall**
(a) light (b) satiate
(c) carry (d) horror

33. **Sacrosanct**
(a) prayer (b) sanctuary
(c) pious (d) sacred

34. **Opprobrium**
(a) mystery (b) opulence
(c) reproach (d) enthusiasm

35. **Judicious**
(a) biased (b) sensible
(c) exact (d) honest

36. **Cogitate**
(a) To live together
(b) To criticize severely
(c) To think deeply about something
(d) To compliment

37. **Vestige**
(a) apology (b) excuse
(c) offence (d) trace

38. **Beaver Away**
(a) To waste away (b) To sleep for long hours
(c) To work hard (d) To seal something

39. **NIP and TUCK**
(a) close competition (b) heavy rain
(c) fierce attack (d) mixture

40. **Adventitious**
(a) accidental
(b) critical
(c) a process using high technology
(d) exciting

41. **Alacrity**
(a) cleanliness (b) leverness
(c) eagerness (d) reluctance

42. **Obstreperous**
(a) fathomless (b) unruly
(c) anathema (d) causing fatigue

43. **Exaction**
(a) demand (b) aggravation
(c) accuracy (d) collaboration

44. **Abominable**
(a) disgusting (b) lovable
(c) abusive (d) undisciplined

45. **Portentous**
(a) imaginary (b) ominous
(c) demonstrative (d) convincing

PRACTICE EXERCISE 2

Which of these words most nearly means the word provided?

1. **Deplete**
(a) decorate (b) beg
(c) exhaust (d) hurry

2. **Voluntary**
(a) willing (b) charity
(c) prisoner (d) careless

3. **Refute**
(a) garbage (b) deny
(c) offer (d) difficult

4. **Cheat**
(a) stingy (b) argue
(c) freckle (d) defraud

5. **Miserable**
(a) cruel (b) wrong
(c) unhappy (d) miss

6. **Vintage**
(a) classic (b) alcoholic
(c) disease (d) spoiled

7. **Tart**
(a) law (b) acid
(c) angry (d) desirable

8. **Corner**
(a) display (b) trap
(c) paint (d) hurry

9. **Zest**
(a) gusto (b) cram
(c) worry (d) trial

10. **Haggle**
(a) tired (b) climb
(c) decrease (d) bargain

11. **Impel**
(a) force (b) block
(c) hinder (d) discredit

12. **Throng**
(a) garment (b) bell
(c) mass (d) weight

13. **Imperial**
(a) bratty (b) oppressive
(c) regal (d) beautiful

14. **Diffuse**
(a) difficult (b) scatter
(c) incomprehensible (d) unplug

15. **Hinder**
(a) lose (b) loose
(c) despair (d) check

16. **Latent**
(a) dormant (b) recent
(c) effeminate (d) desirable

17. **Wretched**
(a) twisted (b) forced
(c) miserable (d) increased

18. **Irksome**
(a) outrageous (b) fearsome
(c) impoverished (d) annoying

19. **Regulate**
(a) even (b) police
(c) flow (d) position

20. Warrant
(a) justify (b) burrow
(c) hide (d) integrity

21. Protract
(a) hire (b) fold
(c) delay (d) corner

22. Lax
(a) ensure (b) slack
(c) servant (d) strive

23. Rigor
(a) austerity (b) rope
(c) fix (d) excess

24. Discrete
(a) leave (b) diminish
(c) squander (d) distinct

25. Lissom
(a) slow (b) honest
(c) supple (d) dull

26. Misprize
(a) despise (b) devalue
(c) erroneous (d) covet

27. Impugn
(a) imply (b) fret
(c) assail (d) recalcitrant

28. Supervene
(a) intervene (b) overreach
(c) displace (d) follow

29. Exigent
(a) urgent (b) treatise
(c) miser (d) expedient

30. Fervid
(a) delightful (b) difficult
(c) obstinate (d) ardent

31. Ersatz
(a) chaotic (b) artificial
(c) impromptu (d) vague

32. Redolent
(a) ubiquitous (b) odorous
(c) shy (d) bellicose

33. Turpitude
(a) lethargy (b) honour
(c) belligerence (d) depravity

34. Propinquity
(a) habit (b) nearness
(c) capacity (d) tendency

35. Vociferous
(a) numerous (b) bountiful
(c) strident (d) garrulous

36. Wraith
(a) Dead body (b) Ghost
(c) Extreme anger (d) Circle of flowers

37. Up the Creek
(a) To get lot of money
(b) In dire difficulties
(c) To be very successful
(d) To achieve one's aim by deceit

38. Obstinate
(a) foolish (b) unyielding
(c) unreasonable (d) mischievous

39. Impasse
(a) deadline (b) stalemate
(c) standstill (d) status quo

40. Abject
(a) miserable (b) objective
(c) tortuous (d) painful

41. Impetus
(a) financial support
(b) accelerated growth
(c) retarded growth
(d) driving force

42. Grotesque
(a) boring (b) gripping
(c) monstrous (d) obnoxious

43. Ephemeral
(a) short-lived (b) immediate
(c) stupendous (d) amazing

44. Derelict
(a) derogatory (b) depressing
(c) ramshackle (d) sluggish

45. Sequestered
(a) secluded
(b) unhealthy
(c) disreputable
(d) marked by affluence

Solutions – PRACTICE EXERCISE – 1

1. **(a) Captious** means prone to fault finding. Synonym: **Fault finding**
2. **(b)** to **abhor** means to feel intense or violent dislike, or to **hate**
3. **(c) graceful** means to be **pleasant** or considerate in social interactions
4. **(a)** Trenchant means strong views; criticism.

   Incisive means showing clear thought and understanding.

   Arduous means difficult to perform.

   Synonym: Incisive
5. **(d) cognate** means people with common ancestors, or **relatives**

6. **(c) meditative** means moodily or dreamily **thoughtful**
7. **(a)** to **expel** means to drive out from home or country, or to **exile**
8. **(c)** a **hoax** is someone who is not what he or she pretends to be, or an **imposter**
9. **(b)** Connubial means related to marriage or conjugal.

    Synonym: Of marriage
10. **(d)** to **haul** is to **pull,** or to cause movement in a direction with applied force
11. **(b) jocund** means good humoured or **merry**
12. **(b)** Pejorative means insulting.

    Disparaging also means insulting or something disapproving.

    Synonym: Disparaging
13. **(b)** to **feign** is to assume the outward appearance of, or to **imitate**
14. **(c) magnetism** is magnetic **charm** or appeal
15. **(a)** to **allocation** is to **divide** and share out
16. **(a) generic** means having the characteristic of a whole group, or **general**
17. **(d)** a **qualm** is a feeling of uneasiness about a moral issue, or a **scruple**
18. **(d) wary** means to be attentive especially to danger, or to be **cautious**
19. **(c)** to **distort** means to twist out of a normal state, or to **deform**
20. **(d) sumptuous** means excessively costly, rich, or **luxurious**
21. **(a)** one meaning of **reel** is to move round and round, or to **whirl**
22. **(b) inscrutable** means not easily interpreted or understood, or **mysterious**
23. **(c)** to **appal** is to overcome with shock, or to **dismay**
24. **(a) upright** can mean either **honourable** or vertical; horizontal and supine are both antonyms of upright
25. **(c)** a **reverie** means the state of being lost in thought, or a **daydream**
26. **(a)** Extortionate means too high (used for price).

    Synonym: Excessive
27. **(a) loquacious** means excessively **talkative,** or garrulous
28. **(c)** a **chimera** is a fabrication of the mind, or an **illusion**
29. **(a) temerity** means unreasonable contempt for danger or recklessness, or **audacity**
30. **(b)** to **educe** means to develop something potential or latent; to **elicit** means to draw out something hidden or latent
31. **(a)** a **nabob** is a person of great wealth or importance, or a **bigwig**
32. **(b)** to **pall** can mean to deprive of pleasure in something by **satiating**
33. **(d) sacrosanct** means the most **sacred,** or holy
34. **(c) opprobrium** means disgrace; public disgrace as a result of bad behaviour.

    Mystery means something difficult to understand.

    Opulence means rich and luxurious; wealthy.

    Reproach means blame; disapproval or disappointment.

    Enthusiasm means excitement.

    Synonym: Reproach (most close)
35. **(b) judicious** means having, showing or done with good sense.
36. **(c) cogitate** means to give careful thought to something; ponder.

    Synonym: To think deeply about something
37. **(d) vestige** means trace of something that once existed.

    Synonym: Trace
38. **(c) beaver away** is a phrasal verb. It means to work very hard at something.

    Synonym: To work hard
39. **(a) Nip and Tuck** (idiom) means to level somebody in a race or competition.

    Synonym: Close competition
40. **(a) adventitious** means accidental; happening by chance.

    Synonym: Accidental
41. **(c) alacrity** means great willingness or enthusiasm.

    Synonym: Eagerness

42. **(b) obstreperous** means noisy; unruly.
Fathomless means easily understood.
Anathema means curse.
Synonym: Unruly

43. **(a) exaction** means demand.
Collaboration means the act of working with another person or group of people to create or produce something.
Synonym: Demand

44. **(a) abominable** means extremely unpleasant and causing disgust.
Synonym: Disgusting; detestable

45. **(b)** portentous means important as a sign or a warning of something that is going to happen in the future, especially when it is something unpleasant.
Synonym: Ominous

Solutions – PRACTICE EXERCISE – 2

1. **(c) deplete** means to reduce or deprive or something essential; **exhaust** means to empty completely
2. **(a) voluntary** means done by one's own will, or **willing**
3. **(b)** to **refute** means to prove wrong, or to **deny** the truth of
4. **(d)** to **cheat** means to influence by means of trickery, or to **defraud**
5. **(c) miserable** means in a state of distress or **unhappiness**
6. **(a) vintage** means of old and enduring interest, or **classic**
7. **(b) tart** means pleasantly sharp or **acid** to the taste
8. **(b)** to **corner** means to drive into a corner, or to **trap**
9. **(a)** one meaning of **zest** is keen enjoyment, or **gusto**
10. **(d)** to **haggle** means to negotiate over terms or price, or to **bargain**
11. **(a)** to **impel** means to drive forward using strong moral pressure, or to **force**
12. **(c)** a **throng** is a large number of assembled people, or a **mass**
13. **(c) imperial** means befitting or suggesting an emperor; **regal** means befitting or suggesting a king
14. **(b)** to **diffuse** means to break up or spread out, or to **scatter**
15. **(d)** to **hinder** means to hold back; one meaning of to **check** means to slow or bring to a stop
16. **(a) latent** means capable of becoming but not currently visible, or **dormant**
17. **(c) wretched** means extremely distressed, or **miserable**
18. **(d) irksome** means tedious or **annoying**
19. **(b)** to **regulate** means to bring under the control of law; to **police** means to control or keep order
20. **(a)** to **warrant** means to serve as adequate ground or reason, or to **justify**
21. **(c)** to **protract** means to prolong in time or space, or to **delay**
22. **(b) lax** means in a relaxed state, or **slack**
23. **(a) rigor** means severity of life, or **austerity**
24. **(d) discrete** means individually **distinct**
25. **(c) lissom** means **supple** or flexible
26. **(a)** to **misprize** means to hold in contempt, or to **despise**
27. **(c)** to **impugn** means to attack verbally as false or lacking integrity; to **assail** means to attack
28. **(d)** to **supervene** means to **follow** as an unexpected development
29. **(a) exigent** means demanding immediate attention, or **urgent**
30. **(d) fervid** means **ardent** or passionate
31. **(b) ersatz** means a usually **artificial** or inferior substitute
32. **(b) redolent** means aromatic or full of a specific scent, or **odorous**

33. **(d) turpitude** means wickedness, or **depravity**
34. **(b) propinquity** means **nearness** in place or time
35. **(c) vociferous** means loud and insistent, often in presentation of demands or requests; **strident** also means loud and insistent
36. **(b)** wraith means the ghost of a person that is seen a short time before or after that person dies.
    Synonym: Ghost
37. **(b) up the creek** (idiom) means in a difficult, unfortunate or extricable position.
    Synonym: In dire difficulties
38. **(b) obstinate** means strubbornly adhering to one's opinion,; not yielding to argument.
39. **(b) impasse** means stalemate or dead lock.
40. **(a) abject** means utterly hopeless, miserable.
41. **(d) impetus** means a moving force, impulse or stimulus.
42. **(c) grotesque** means odd or unnatural in shape or size.
43. **(a) ephemeral** means shortlived, transitory.
44. **(c) derelict** means abandoned, deserted, bedraggled, ramshackle.
45. **(a) sequestered** means quiet and far away from people.
    Synonym: Secluded

ANTONYM

Introduction

Questions based on "Antonyms" are certainly an important section of General English. This is a test in which your ability to understand the meanings of words and to distinguish between the fine shades of meaning is checked. While testing your ability on questions pertaining to Antonyms, the first thing to be observed is **your ability to grasp the meaning of the words given**. Unless you know the meaning of the words given, you will not be able to find out or choose the exact antonyms from the options given. Mostly antonyms appear in the form of nouns, verbs and Adjectives.

EXAMPLES —

Fortunate - Unfortunate

Satisfied - Dissatisfied

Complete - Incomplete

Polite - Impolite

Similarly, the suffixes can also give opposite meanings. For examples Less is a suffix which gives an opposite meaning.

EXAMPLES —

Hope - Hopeless

Meaning - Meaningless

Tips & Tricks for SOLVE QUESTIONS BASED ON ANTONYMS

Your approach to solve the questions based on antonyms require some strategies.

- Check whether the question word and the opposite of the words given under the options are in the **same parts of speech** i.e. noun, verb to verb etc.

- You should have an idea of **the roots of the words and know their meanings.**

Example ▶ The words "benefactor", if you know the meaning of "bene" you will be able to guess the meanings of the words, and opt for the best antonyms.

- You should also have the knowledge of **prefixes and suffixes** this will help you to find out the suitable antonyms.

Example ▶ The prefix "un", "in", "im", "dis" etc. give negative meanings.

- **Some words may not have a precise antonym. In such cases, you can look for a word or a phrase which is nearly the opposite.**
- Sometimes, note than one word in the options given word. So, one has to make the choice judiciously. In this case, choose the word which gives great detail in the meaning.
- Looking for the best answer and not for the ideas answer. Eliminate two or three of the options if they are nowhere related to given word.
- Do not go for an antonym which is too limited or too broad to be an opposite.
- when you get confused about the antonym to be chosen, try to think of how you have heard the word used before. You may discover a suitable context to guess the exact antonym.

Some more examples solved with the suggested approach.

1. **DILAPIDATED**
   (a) ruined (b) condemned
   (c) renovated (d) destroyed

"Dilapidated" means falling apart therefore options (a) and (d) can be eliminated as they show intentional actions. The only word, meaning made good as new, is renovated" it standout as the best antonym. Options (b) means" to disapprove" so that is also not suitable.

2. **MELODIOUS**
   (a) mellifluous (b) unpleasant
   (c) spiritual (d) comfort

In this example, the word "Melodious" means "Mellifluous, which in turn means pleasant in sound the options (c) and (d) are irrelevant in this context the only antonym is "b"- Unpleasant.

3. **PROFUSION**
   (a) Deficiency (b) Certainty
   (c) Protuberating (d) Largeness

"Pro" is s Prefix which has an opposite connotation. You can notice "de" in the word "Deficiency" is another prefix which has a negative meaning for example defrost, Dethrone etc.

"Profusion" means abundance" Deficiency means "Inadequacy therefore, the exact opposite of profusion.

PRACTICE EXERCISE

Directions:
Each of the sample antonyms questions consists of a word followed by five words or phrase as choices. Choose the word or phrase which is most nearly opposite in meaning to the word in capitals and shade the alphabets marked in the grid on your answer sheet.

Following are some sample antonyms questions based on the entrance tests of various PSUs.

1. **EXECRATE**
   (a) loathe (b) despise
   (c) adore (d) abhor
2. **SYCOPHANT**
   (a) servile (b) first
   (c) fawning (d) supercilious
3. **OSTENTATIOUS**
   (a) not resonant (b) not reddish
   (c) resonant (d) pompous
4. **RETRACT**
   (a) entangle (b) rescue
   (c) fail (d) assert
5. **REPROACH**
   (a) defer (b) vacillate
   (c) sever (d) laud
6. **PLENTIFULNESS**
   (a) luxury (b) magnificence
   (c) scarcity (d) contentment
7. **ABUSIVE**
   (a) decent (b) savage
   (c) major (d) volatile
8. **ALTERCATION**
   (a) accord (b) repetition
   (c) escape (d) ratification
9. **DILATE**
   (a) deteriorate (b) weaken
   (c) constrict (d) concentrate
10. **TOUT**
    (a) cast aspersions on
    (b) misrepresent
    (c) placate
    (d) withhold consent
11. **SQUALID**
    (a) fervid (b) florid
    (c) pristine (d) extraneous

12. INVECTIVE
(a) willing compliance
(b) normality
(c) restoration
(d) laudatory words

13. PERFIDY
(a) tact (b) generosity
(c) thoroughness (d) loyalty

14. OUTLANDISH
(a) conventional (b) prolific
(c) unchanging (d) transparent

15. PLUMB
(a) reversed (b) lofty
(c) horizontal (d) thin

16. FERVID
(a) ardent (b) unexpected
(c) stubborn (d) restrained

17. VACUITY
(a) quality (b) certainty
(c) plenitude (d) stability

18. RAVEL
(a) knit (b) omit
(c) remain silent (d) measure

19. PERSISTENCE
(a) irrelevance (b) cessation
(c) inequality (d) intemperance

20. SUB ROSA
(a) openly (b) fashionably
(c) clandestinely (d) simply

21. INVIDIOUS
(a) candid (b) stubborn
(c) defensive (d) inoffensive

22. TURPITUDE
(a) pragmatism (b) probity
(c) judiciousness (d) animation

23. SKEPTICISM
(a) reason (b) suspicion
(c) plausibility (d) audacity

24. IGNOMINIOUS
(a) scholarly (b) incognito
(c) laudatory (d) disgraceful

25. CALUMINATE
(a) accuse (b) libel
(c) laud (d) asperse

26. PALTRY
(a) plenty (b) scanty
(c) petty (d) friendly

27. VULNERABLE
(a) frantic (b) feeble
(c) secure (d) complicated

28. BRITTLE
(a) thin (b) breakable
(c) grating (d) flexible

29. DAINTY
(a) delicate (b) coarse
(c) harse (d) delicious

30. INNOCUOUS
(a) safe (b) injurious
(c) sober (d) impeccable

31. ANIMOSITY
(a) camaraderie (b) retardation
(c) sincerity (d) refutation

32. INVETERATE
(a) uninvited (b) addicted
(c) inbred (d) beginning

33. SCOTCH
(a) renovate (b) generous
(c) entrust (d) ameliorate

34. PREDILECTION
(a) ambiguity
(b) unwillingness to choose
(c) desire to please
(d) propensity to dislike

35. CHOLERIC
(a) good-natured (b) spoiled
(c) irascible (d) feisty

36. EXACERBATE
(a) contemplate (b) bewilder
(c) worsen (d) appease

37. EQUANIMITY
(a) clamour (b) volume
(c) disparity (d) agitation

38. VALIANT
(a) instinctive (b) cynical
(c) cowardly (d) worthy

39. EXHUME
(a) enter (b) fertilize
(c) inter (d) decay

40. CALLOW
(a) rustic (b) crude
(c) exquisite (d) experienced

41. CUPIDITY
(a) generosity (b) love
(c) anxiety (d) entertainment

42. SUCCINCT
(a) distinct (b) verbose
(c) feeble (d) iaconic

43. UNTOWARD
(a) experienced (b) inevitable
(c) industrious (d) favourable

44. SATURNINE
(a) festive (b) genial
(c) heavy handed (d) distinguished

45. PERSPICACIOUS
(a) Insufficiently precise
(b) of indefinite duration
(c) dull wilted
(d) penetrating

46. INCARCERATE
(a) summon (b) assist
(c) liberate (d) confine

47. INSOLVENCY
(a) ability to sustain growth
(b) concentration
(c) coherence
(d) ability to pay one's debts

48. NONCHALANT
(a) intelligent (b) popular
(c) concerned (d) reckless

49. APPOSITE
(a) parallel (b) synonymous
(c) hostile (d) inappropriate

50. GRATUITOUS
(a) frank (b) pithy
(c) warranted (d) frugal

51. PREFATORY
(a) intelligent (b) outstanding
(c) predatory (d) conclusive

52. CONCILIATE
(a) arrive (b) appeal
(c) incite (d) estrange

53. SUBSERVIENT
(a) fawning (b) obsequious
(c) domineering (d) omnipresent

54. VAUNTED
(a) berated (b) belittled
(c) lauded (d) wicked

55. QUOTA
(a) Anonymous remark
(b) decisive action
(c) debatable issue
(d) unlimited number

56. CONTENTIOUS
(a) satisfied (b) pacific
(c) hungry (d) bellicose

57. OBLOQUY
(a) fame (b) censure
(c) inquiry (d) exaltation

58. PENCHANT
(a) distaste (b) scabbard
(c) agreement (d) itch

59. BALEFUL
(a) empty (b) menacing
(c) gaudy (d) congenial

60. BAROQUE
(a) rococo (b) simple
(c) bizarre (d) ornate

61. MYOPIC
(a) blind (b) moral
(c) visionary (d) presbyopic

62. NASCENT
(a) germinal (b) fading
(c) unnamed (d) treacherous

63. DISSONANCE
(a) harmony (b) carefulness
(c) specificity (d) value

64. TURBULENCE
(a) immunity (b) tranquillity
(c) meditation (d) coordination

65. BANAL
(a) inclined (b) faithful
(c) elaborate (d) forced

66. GERMINAL
(a) sterilized (b) sprouting
(c) fully developed (d) primitive

67. ERUDITE
(a) uneducated (b) polite
(c) unknown (d) agitated

68. INFANTILE
(a) despicable (b) mature
(c) adolescent (d) perpetual

69. OPPORTUNIST
(a) expedient (b) fatalist
(c) fledgling (d) inexpedient

70. CENSURE
(a) reprimand (b) encomium
(c) criticism (d) legal construal

71. COMMODIOUS
(a) product (b) space
(c) cramped (d) company

72. EFFRONTERY
(a) modesty (b) diffidence
(c) avoidance (d) shamelessness

73. OBSTREPEROUS
(a) noisy (b) defiant
(c) permeable (d) quiet

74. PACIFY
(a) ameliorate (b) patch up
(c) agitated (d) tormented

75. AMBIGUOUS
(a) confusing (b) lucid
(c) desirous (d) obfuscate

76. MILITANT
(a) Dramatic (b) combative
(c) religious (d) pacific

77. MOTILITY
(a) apathy (b) tension
(c) ascent (d) preceiving

78. SINUOUS
(a) wet (b) vacant
(c) numerous (d) direct

79. PRIM
(a) last (b) outrageous
(c) ugly (d) cantankerous

80. SEGMENT
(a) inflate (b) affix
(c) keep still (d) make whole

81. OMNISCIENT
(a) subordinate (b) ignorant
(c) ignoble (d) weak

82. JEOPARDISE
(a) help (b) encourage
(c) take care (d) preserve

83. ADVERSARY
(a) acquaintance (b) friend
(c) enemy (d) competitor

84. CONCEIT
(a) humility (b) pride
(c) determination (d) arrogance

85. IMPUDENCE
(a) tolerance (b) modesty
(c) patience (d) prudence

86. SALACIOUS
(a) decent (b) satisfying
(c) precipitous (d) fortifying

87. ESTRANGEMENT
(a) reunion (b) familiarity
(c) gathering (d) meeting

88. LUKEWARM
(a) enthusiastic (b) moderate
(c) friendly (d) fervent

89. RUTHLESS
(a) benevolent (b) obliging
(c) merciful (d) affectionate

90. ARRAIGN
(a) punish (b) pardon
(c) summon (d) indict

Answer Key

1	(c)	11	(c)	21	(d)	31	(a)	41	(a)	51	(d)	61	(c)	71	(c)	81	(b)
2	(d)	12	(d)	22	(b)	32	(d)	42	(b)	52	(c)	62	(b)	72	(b)	82	(d)
3	(a)	13	(d)	23	(d)	33	(b)	43	(d)	53	(c)	63	(a)	73	(d)	83	(c)
4	(d)	14	(a)	24	(b)	34	(d)	44	(a)	54	(c)	64	(d)	74	(c)	84	(a)
5	(d)	15	(c)	25	(c)	35	(a)	45	(c)	55	(d)	65	(c)	75	(b)	85	(d)
6	(c)	16	(d)	26	(a)	36	(d)	46	(c)	56	(a)	66	(d)	76	(d)	86	(a)
7	(a)	17	(c)	27	(c)	37	(d)	47	(d)	57	(d)	67	(a)	77	(a)	87	(a)
8	(a)	18	(a)	28	(a)	38	(c)	48	(c)	58	(a)	68	(b)	78	(d)	88	(a)
9	(c)	19	(b)	29	(b)	39	(c)	49	(d)	59	(d)	69	(d)	79	(b)	89	(c)
10	(b)	20	(a)	30	(b)	40	(d)	50	(c)	60	(b)	70	(b)	80	(d)	90	(b)

CHAPTER 5 One Word Substitutions

Set-1 Out of the **four alternatives** choose the one which can be substituted for the given words/sentences:

1. Person leading a life of strict self discipline.
(a) hedonist (b) disciplination
(c) disciplinarian (d) ascetic

2. A person who loves everybody.
(a) egoist (b) fatalist
(c) humanist (d) altruist

3. A government run by officials.
(a) democracy (b) anarchy
(c) oligarchy (d) bureaucracy

4. That which can be drink.
(a) edible (b) palatable
(c) potable (d) culpable

5. The worship of idols or images.
(a) atheism (b) theism
(c) idolatry (d) iconoclasm

6. Something that is poisonous or unhealthy.
(a) trivial (b) toxic
(c) torpid (d) tragic

7. A remedy for all disease.
(a) amnesia (b) panacea
(c) intelligentia (d) parasol

8. A hater of mankind.
(a) misanthrope (b) misogynist
(c) philanthropist (d) misogamist

9. General view of a person's character.
(a) biography (b) profile
(c) introduction (d) sketch

10. Placing a thing beside another.
(a) impose (b) repose
(c) juxtapose (d) expose

11. The habit of always admiring onself.
(a) psychosis (b) Neurosis
(c) narcissism (d) paranoia

12. A pan for small animals.
(a) hutch (b) lair
(c) den (d) cage

13. A woman having more than one husband at the same time.
(a) polyandry (b) polyphony
(c) polygamy (d) polygyny

14. A group of birds.
(a) bevy (b) covey
(c) troupe (d) coterie

15. Causing in ending death.
(a) fatal (b) deadly
(c) serious (d) dangerous

16. One who loves books.
(a) scholar (b) bibliographer
(c) teacher (d) bibliophile

17. A supporter of the cause of women.
(a) effeminate (b) loquacious
(c) sophist (d) feminist

18. That which is contrary to law.
(a) legal (b) legacy
(c) illegal (d) imminent

19. One who comes to settle in a country.
(a) tourist (b) emigrant
(c) immigrant (d) visitor

20. A murderer of one's father.
(a) fratricide (b) homicide
(c) patricide (d) regicide

21. One who flirts with ladies.
(a) solvent (b) gentleman
(c) popular (d) philanderer

22. One that lives on another.
(a) crazy (b) parasite
(c) intelligent (d) orphan

23. A person who practices celibacy.
(a) unmarried (b) yogi
(c) celibate (d) devil

24. Morbid fear of confined spaces.
(a) dipsomania (b) xenophobia
(c) claustrophobia (d) pyromania

25. A language which is no longer is use.
(a) obsolete (b) dead language
(c) phonetics (d) morphology

Set-2 Out of the **four alternatives** choose the one which can be substituted for the given words/sentences:

1. An allowance made to a wife by her husband, whey they are legally separated.
(a) alimony (b) parsimony
(c) matrimony (d) honorarium

2. Fear of telling lies.
(a) mythophobia (b) legendary
(c) lygophobia (d) wagoner

3. Walking in sleep,
(a) obsession (b) hallucination
(c) somnambulism (d) somniloquism

4. A particular method of working.
(a) method (b) knack
(c) working pattern (d) modus operandi

5. A place where dead bodies are buried.
(a) museum (b) cemetery
(c) crematory (d) park

6. Medicine given to counteract poison.
(a) antibiotic (b) antiseptic
(c) antidote (d) antifungal

7. Non likely to be easily pleased.
(a) fastidious (b) infallible
(c) fatalist (d) communist

8. Shaking moment of the ground.
(a) tremor (b) shiver
(c) vibrate (d) oscillate

9. Liable to be easily broken.
(a) translucent (b) brittle
(c) unhealthy (d) opaque

10. Easy to shape in any desired form.
(a) ductile (b) malleable
(c) flexible (d) plasticine

11. Amount of money demanded by kidnappers.
(a) donation (b) honorarium
(c) ransom (d) salary

12. Strong disliking between two people.
(a) aversion (b) antipathy
(c) apathy (d) despair

13. A place where wine is made.
(a) bakery (b) clockroom
(c) tannery (d) brewery

14. Found all over the world.
(a) international (b) universal
(c) regional (d) provincial

15. A post with a little work, but high salary.
(a) director (b) trustee
(c) sinecure (d) ombudsman

16. Occurring at night.
(a) nighty (b) dark
(c) neurotic (d) noctural

17. Careful and thorough enquiry.
(a) investigation (b) interview
(c) examination (d) exploration

18. Seeing something which is not actually present.
(a) imagination (b) presumption
(c) hallucination (d) supposition

19. Science of heredity.
(a) hereditary (b) genetics
(c) genesis (d) inheritance

20. A specialist who tests eyesight.
(a) optician (b) ophthalmologist
(c) ichthyologist (d) neurologist

21. A person who has no money to pay off his debts.
(a) Insolvent (b) poor
(c) destitute (d) pauper

22. A person who breaks into house in order to steal.
(a) poacher (b) bandit
(c) intruder (d) burglar

23. Tough tissues in joints.
(a) ligaments (b) endoderm
(c) muscles (d) fibre

24. The first model of a new device.
(a) prototype (b) sculpture
(c) icon (d) photograph

25. Customs and habits of a particular group.
(a) mores (b) traditions
(c) rites (d) rituals

Set-3 Out of the **four alternatives** choose the one which can be substituted for the given words/sentences:

1. A woman who remains unmarried.
(a) celibate (b) spinster
(c) virgin (d) stunning

2. The examination of a dead body to find the cause of death.
(a) autopsy (b) bellicose
(c) postnatal (d) biopsy

3. One who goes on foot.
(a) commuter (b) pedestrian
(c) stoic (d) scholar

4. One who eats human flesh.
(a) hypocrite
(b) non-vegetarian
(c) cannibal
(d) eggetarian

5. A list of books.
(a) library (b) menu
(c) catalogue (d) bookstore

6. A person who is unaffected by emotions.
(a) stoic (b) celibate
(c) handsome (d) curator

7. Any kind of grain used for food.
(a) rice (b) pulses
(c) cereal (d) none of those

8. The art of beautiful handwriting.
(a) calligraphy (b) photography
(c) euphemism (d) carving

9. One who deserts his principle or beliefs.
(a) renegade (b) recant
(c) convent (d) affable

10. A place where nuns live and work.
(a) Convent (b) Church
(c) den (d) holyroom

11. A place where birds are kept
(a) aviary (b) house
(c) nest (d) hut

12. The act of speaking about one's thought when one is alone.
(a) Sonnet (b) ballad
(c) soliloquy (d) elegy

13. One who helps people by giving them money or other aid.
(a) benefactor (b) well wisher
(c) parents (d) advocate

14. An inscription on a tomb
(a) notorious (b) patrimony
(c) epitaph (d) pseudonym

15. A speech delivered without any preparation.
(a) extempore (b) nomads
(c) debate (d) conversation

16. One who deals in flowers.
(a) a florist (b) coy
(c) diffident (d) feminist

17. That which cannot be defeated.
(a) fastidious (b) warden
(c) invincible (d) effeminate

18. An assembly of worshipper.
(a) congregation (b) convent
(c) usurer (d) congress

19. A person who abstains completely from alcoholic drinks.
(a) versatile (b) plaintiff
(c) teetotaller (d) drunkard

20. A list of luggage.
(a) waybill (b) horizon
(c) catalogue (d) menu

21. A set of three related works by the same author.
(a) trilogy (b) morphology
(c) cartography (d) choreography

22. A short stay at a place.
(a) solvent (b) rest
(c) sojourn (d) impunity

23. A dramatic entertainment, in which performers express through gestures.
(a) pantomime (b) decamp
(c) mandible (d) none of these

24. Comment added to a text or diagram.
(a) example (b) note
(c) annotation (d) depict

25. Greedy for money
(a) rapacious (b) ravenous
(c) miser (d) coronation

Answer Key : Set-1 One Word Substitution

1. (c)	2. (d)	3. (d)	4. (c)	5. (c)	6. (b)
7. (b)	8. (a)	9. (b)	10. (c)	11. (c)	12. (a)
13. (c)	14. (b)	15. (a)	16. (b)	17. (d)	18. (c)
19. (c)	20. (c)	21. (d)	22. (b)	23. (c)	24. (c)
25. (b)					

Answer Key : Set-2 One Word Substitution

1. (a)	2. (a)	3. (c)	4. (d)	5. (b)	6. (c)
7. (a)	8. (a)	9. (b)	10. (c)	11. (c)	12. (b)
13. (d)	14. (b)	15. (c)	16. (d)	17. (a)	18. (c)
19. (b)	20. (b)	21. (a)	22. (d)	23. (a)	24. (a)
25. (a)					

Answer Key : Set-3 One Word Substitution

1. (b)	2. (a)	3. (b)	4. (c)	5. (c)	6. (a)
7. (c)	8. (a)	9. (b)	10. (a)	11. (a)	12. (c)
13. (a)	14. (c)	15. (a)	16. (a)	17. (c)	18. (a)
19. (c)	20. (a)	21. (a)	22. (c)	23. (a)	24. (c)
25. (a)					

Section

C

GATE

Previous Years Solved Questions

1. Which of the following options is the closest in meaning to the word below?
Circuitous
(a) Cyclic (b) Indirect
(c) Confusing (d) Crooked
[2010, 1 Mark]

2. The question below consists of a pair of related words followed by four pairs of words. Select the pair that best expresses the relation in the original pair.
Unemployed: Worker
(a) Fallow : Land
(b) Unaware : Sleeper
(c) Wit : Jester
(d) Renovated : House
[2010, 1 Mark]

3. Choose the most appropriate word from the options given below to complete the following sentence:
If we manage to .... our natural resources, we would leave a better planet for our children.
(a) uphold (b) restrain
(c) cherish (d) conserve
[2010, 1 Mark]

4. Choose the most appropriate word from the options given below to complete the following sentence:
His rather casual remarks on politics .... his lack of seriousness about the subject.
(a) masked (b) belied
(c) betrayed (d) suppressed
[2010, 1 Mark]

5. Modern warfare has changed from large scale clashes of armies to suppression of civilian populations. Chemical agents that do their work silently appear to be suited to such warfare; and regretfully, there exist people in military establishments who think that chemical agents are useful fools for their cause.
Which of the following statements best sums up the meaning of the above passage?
(a) Modern warfare has resulted in civil strife.
(b) Chemical agents are useful in modern warfare.
(c) Use of chemical agents in warfare would be undesirable.
(d) People in military establishments like to use chemical agents in war.
[2010, 2 Marks]

6. Choose the word from the options given below that is most nearly opposite in meaning to the given word:
Amalgamate
(a) merge (b) split
(c) collect (d) separate
[CE, ME, CS, 2011, 1 Mark (Set-1)]

7. Which of the following options is the closest in the meaning to the word below:
Inexplicable
(a) Incomprehensible (b) Indelible
(c) Inextricable (d) Infallible
[CE, ME, CS, 2011, 1 Mark (Set-1)]

8. Choose the most appropriate word(s) from the options given below to complete the following sentence.
I contemplated__________ Singapore for my vacation but decided against it.
(a) to visit
(b) having to visit
(c) visiting
(d) for a visit
[CE, ME, CS, 2011, 1 Mark (Set-1)]

9. Choose the most appropriate word from the options given below to complete the following sentence.
If you are trying to make a strong impression on your audience, you cannot do so by being understated, tentative or
(a) hyperbolic (b) restrained
(c) argumentative (d) indifferent
[CE, ME, CS, 2011, 1 Mark (Set-1)]

10. **Few school curricula include a unit on how to deal with bereavement and grief, and yet all students at some point in their lives suffer from losses through death and parting.**
Based on the above passage which topic would not be included in a unit on bereavement?
(a) how to write a letter of condolence
(b) what emotional stages are passed through in the healing process
(c) what the leading causes of death are
(d) how to give support to a grieving friend
[CE, ME, CS, 2011, 2 Marks (Set-1)]

11. The question below consists of a pair of related words followed by four pairs of words. Select the pair that best expresses the relation in the original pair:
Gladiator : Arena
(a) dancer : stage
(b) commuter : train
(c) teacher : classroom
(d) lawyer : courtroom
[EE, EC, 2011, 1 Mark (Set-2)]

12. Choose the most appropriate word from the options given below to complete the following sentence:
It was her view that the country's problems had been ________ by foreign technocrats, so that to invite them to come back would be counter-productive.
(a) identified (b) ascertained
(c) exacerbated (d) analysed
[EE, EC, 2011, 1 Mark (Set-2)]

13. Choose the word from the options given below that is most nearly opposite in meaning to the given word:
Frequency
(a) periodicity (b) rarity
(c) gradualness (d) persistency
[EE, EC, 2011, 1 Mark (Set-2)]

14. Choose the most appropriate word from the options given below to complete the following sentence:
Under ethical guidelines recently adopted by the Indian Medical Association, human genes are to be manipulated only to correct diseases for which ______ treatments are unsatisfactory.
(a) similar (b) most
(c) uncommon (d) available
[EE, EC, 2011, 1 Mark (Set-2)]

15. The horse has played a little known but very important role in the field of medicine. Horses were injected with toxins of diseases until their blood built up immunities. Then a serum was made from their blood. Serums to fight with diphtheria and tetanus were developed this way.
It can be inferred from the passage, that horses were
(a) given immunity to diseases
(b) generally quite immune to diseases
(c) given medicines to fight toxins
(d) given diphtheria and tetanus serums
[EE, EC, 2011, 2 Marks (Set-2)]

16. Choose the most appropriate alternative from the options given below to complete the following sentence:
Despite several ______ the mission succeeded in its attempt to resolve the conflict.
(a) attempts (b) setbacks
(c) meetings (d) delegations
[CE, ME CS, 2012, 1 Mark (Set-1)]

17. Which one of the following options is the closest in meaning to the word given below?
Mitigate
(a) Diminish (b) Divulge
(c) Dedicate (d) Denote
[CE, ME CS, 2012, 1 Mark (Set-1)]

18. Choose the grammatically INCORRECT sentence:
(a) They gave us the money back less the service charges to three Hundred rupees.
(b) This country's expenditure is not less than that of Bangladesh.
(c) The committee initially asked for a funding of fifty lakh rupees, but later settled for a lesser sum.
(d) This country's expenditure on educational reforms is very less.
[CE, ME CS, 2012, 1 Mark (Set-21]

19. Choose the most appropriate alternative from the options given below to complete the following sentence:
Suresh's dog is the one _____ was hurt in the stampede.
(a) that (b) which
(c) who (d) whom
[CE, ME CS, 2012, 1 Mark (Set-1)]

20. Wanted Temporary, Part-time persons for the post of Field Interviewer to conduct personal interviews to collect and collate economic data. Requirements: High School-pass, must be available for Day, Evening and Saturday work. Transportation paid, expenses reimbursed.
Which one of the following is the best inference from the above advertisement?

(a) Gender-discriminatory
(b) Xenophobic
(c) Not designed to make the post attractive
(d) Not gender-discriminatory

[CE, ME CS, 2012, 2 Marks (Set-1)]

21. Choose the most appropriate alternative from the options given below to complete the following sentence:

If the tired soldier wanted to lie down, he ____ the mattress out on the balcony.

(a) should take
(b) shall take
(c) should have taken
(d) will have taken

[EE, EC, 2012, 1 Mark (Set-2)]

22. One of the parts (A, B, C, D) in the sentence given below contains an ERROR. Which one of the following is INCORRECT?

I requested that he should be given the driving test today instead of tomorrow.

(a) requested that
(b) should be given
(c) the driving test
(d) instead of tomorrow

[EE, EC, 2012, 1 Mark (Set-2)]

23. Which one of the following options is the closest in meaning to the word given below?

Latitude

(a) Eligibility (b) Freedom
(c) Coercion (d) Meticulousness

[EE, EC, 2012, 1 Mark (Set-2)]

24. Choose the most appropriate word from the options given below to complete the following sentence:

Given the seriousness of the situation that he had to face, his ___ was impressive.

(a) beggary (b) nomenclature
(c) jealousy (d) nonchalance

[EE, EC, 2012, 1 Mark (Set-2)]

25. One of the legacies of the Roman legions was discipline. In the legions, military law prevailed and discipline was brutal. Discipline on the battlefield kept units obedient, intact and fighting, even when the odds and conditions were against them.

Which one of the following statements best sums up the meaning of the above passage?

(a) Thorough regimentation was the main reason for the efficiency of the Roman legions even in adverse circumstances.
(b) The legions were treated inhumanly as if the men were animals.
(c) Discipline was the armies inheritance from their seniors.
(d) The harsh discipline to which the legions were subjected to led to the odds and conditions being against them.

[EE, EC, 2012, 2 Marks (Set-2)]

26. Friendship, No matter how ____ it is, has its limitation

(a) cordial (b) intimate
(c) secret (d) pleasant

[CE, 2013, 1 Mark (Set-1)]

27. The pair that best express a relationship similar to that expression in the pair:

Medicine : Health

(a) Science : Experiment
(b) Wealth : Peace
(c) Education : Knowledge
(d) Money : Happiness

[CE, 2013, 1 Mark (Set-1)]

28. Which of the following options is closest in meaning to the word given below:

"Primeval"

(a) Modern (b) Historic
(c) Primitive (d) Antique

[CE, 2013, 1 Mark (Set-1)]

29. The professor (I) ordered to (II) the student to go (III) out of the class (IV)

The incorrect one is

(a) (I) (b) II
(c) (III) (d) (IV)

[CE, 2013, 1 Mark (Set-1)]

30. Abhishek is elder to Savan, Savan is younger to Anshul. The correct relations is

(a) Abhishek is elder to Anshul
(b) Anshul is elder to Abhishek
(c) Abhishek and Anshul are of same age
(d) No conclusion can be drawn

[CE, 2013, 2 Marks (Set-1)]

31. Complete the sentence :
Universalism is to particularism as diffuseness is to ____.
(a) specificity (b) neutrality
(c) generality (d) adaptation
[ME, PI & CS/IT, 2013, 1 Mark (Set-2)]

32. Were you a bird, you _____ in the sky.
(a) would fly (b) shall fly
(c) should fly (d) shall have flown
[ME, PI & CS/IT, 2013, 1 Mark (Set-2)]

33. Which one of the following options is the closest in meaning to the word given below?
Nadir
(a) Highest (b) Lowest
(c) Medium (d) Integration
[ME, PI & CS/IT, 2013, 1 Mark (Set-2)]

34. Choose the grammatically **INCORRECT** sentence:
(a) He is of Asian origin.
(b) They belonged to Africa
(c) She is an European.
(d) They migrated from India to Australia.
[ME, PI & CS/IT, 2013, 1 Mark (Set-2)]

35. After several defeats in wars, Robert Bruce went in exile and wanted to commit suicide. Just before committing suicide, he came across a spider attempting tirelessly to have its net. Time and again, the spider failed but that did not deter it to refrain from making attempts. Such attempts by the spider made Bruce curious. Thus, Bruce started observing the near-impossible goal of the spider to have the net. Ultimately, the spider succeeded in having its net despite several failures. Such act of the spider encouraged Bruce not to commit suicide. And then, Bruce went back again and won many a battle, and the rest is history.
Which one of the following assertions is best supported by the above information?
(a) Failure is the pillar of success
(b) Honesty is the best policy
(c) Life begins and ends with adventures
(d) No adversity justifies giving up hope
[ME, PI & CS/IT, 2013, 2 Marks (Set-2)]

36. Choose the grammatically CORRECT sentence:
(a) Two and two add four.
(b) Two and two become four.
(c) Two and two are four.
(d) Two and two make four.
[EE, EC & IN, 2013, 1 Mark (Set-3)]

37. **Statement:** You can always give me a ring whenever you need.
Which one of the following is the best inference from the above statement?
(a) Because I have a nice caller tune.
(b) Because I have a better telephone facility.
(c) Because a friend in need is a friend indeed.
(d) Because you need not pay towards the telephone bills when you give me a ring.
[EE, EC & IN, 2013, 1 Mark (Set-3)]

38. Complete the sentence:
Dare _______ mistakes.
(a) commit (b) to commit
(c) committed (d) committing
[EE, EC & IN, 2013, 1 Mark (Set-3)]

39. They were requested not to quarrel with others.
Which one of the following options is the closest in meaning to the word **quarrel**?
(a) make out (b) call out
(c) dig out (d) fall out
[EE, EC & IN, 2013, 1 Mark (Set-3)]

40. **Statement:** There were different streams of freedom movements in colonial India carried out by the moderates, liberals, radicals, socialists, and so on.
Which one of the following is the best inference from the above statement?
(a) The emergence of nationalism in colonial India led to our Independence.
(b) Nationalism in India emerged in the context of colonialism.
(c) Nationalism in India is homogeneous.
(d) Nationalism in India is heterogeneous.
[EE, EC & IN, 2013, 2 Marks (Set-3)]

41. Choose the most appropriate phrase from the options given below to complete the following sentence.
The aircraft _______ take off as soon as its flight plan was filed.
(a) is allowed to (b) will be allowed to
(c) was allowed to (d) has been allowed to
[EC & ME, 2014, 1 Mark (Set-1)]

42. Read the statements:
All women are entrepreneurs.
Some women are doctors.
Which of the following conclusions can be logically inferred from the above statements?
(a) All women are doctors
(b) All doctors are entrepreneurs
(c) All entrepreneurs are women
(d) Some entrepreneurs are doctors
[EC & ME, 2014, 1 Mark (Set-1)]

43. Choose the most appropriate word from the options given below to complete the following sentence.
Many ancient cultures attributed disease to supernatural causes. However, modern science has largely helped _____ such notions.
(a) impel (b) dispel
(c) propel (d) repel
[EC & ME, 2014, 1 Mark (Set-1)]

44. Choose the most appropriate word from the options given below to complete the following sentence.
Communication and interpersonal skills are _______ important in their own ways.
(a) each (b) both
(c) all (d) either
[EC & ME, 2014, 1 Mark (Set-2)]

45. Which of the options given below best completes the following sentence?
She will feel much better if she _________.
(a) will get some rest
(b) gets some rest
(c) will be getting some rest
(d) is getting some rest
[EC & ME, 2014, 1 Mark (Set-2)]

46. Choose the most appropriate pair of words from the options given below to complete the following sentence.
She could not _____ the thought of _______ the election to her bitter rival.
(a) bear, loosing (b) bare, loosing
(c) bear, losing (d) bare, losing
[EC & ME, 2014, 1 Mark (Set-2)]

47. "India is a country of rich heritage and cultural diversity." Which one of the following facts best supports the claim made in the above sentence?
(a) India is a union of 28 states and 7 union territories.
(b) India has a population of over 1.1 billion.
(c) India is home to 22 official languages and thousands of dialects.
(d) The Indian cricket team draws players from over ten states.
[EC & ME, 2014, 1 Mark (Set-3)]

48. The value of one U.S. dollar is 65 Indian Rupees today, compared to 60 last year. The Indian Rupee has ___________.
(a) depressed
(b) depreciated
(c) appreciated
(d) stabilized
[EC & ME, 2014, 1 Mark (Set-3)]

49. 'Advice' is ___________.
(a) a verb
(b) a noun
(c) an adjective
(d) both a verb and a noun
[EC & ME, 2014, 1 Mark (Set-3)]

50. Which of the following options is the closest in meaning to the word underlined in the sentence below?
In a democracy, everybody has the freedom to **disagree** with the government.
(a) dissent (b) descent
(c) decent (d) decadent
[EC & ME, 2014, 1 Mark (Set-4)]

51. After the discussion, Tom said to me, 'Please revert!'. He expects me to _________.
(a) retract (b) get back to him
(c) move in reverse (d) retreat
[EC & ME, 2014, 1 Mark (Set-4)]

52. While receiving the award, the scientist said, "I feel vindicated". Which of the following is closest in meaning to the word 'vindicated'?
(a) punished (b) substantiated
(c) appreciated (d) chastened
[EC & ME, 2014, 1 Mark (Set-4)]

53. Which of the following options is the closest in meaning to the phrase underlined in the sentence below?

It is fascinating to see life forms <u>cope with</u> varied environment conditions.

(a) adopt to (b) adapt to
(c) adept in (d) accept with

[EE & CS, 2014, 1 Mark (Set-1)]

54. Choose the most appropriate word from the options given below to complete the following sentence.

He could not understand the judges awarding her the first prize, because he thought that her performance was quite ______.

(a) superb (b) medium
(c) mediocre (d) exhilarating

[EE & CS, 2014, 1 Mark (Set-1)]

55. In a press meet on the recent scam, the minister said, "The buck stops here". What did the minister convey by the statement?

(a) He wants all the money
(b) He will return the money
(c) He will assume final responsibility
(d) He will resist all enquiries

[EE & CS, 2014, 1 Mark (Set-1)]

56. The Palghat Gap (or Palakkad Gap), a region about 30 km wide in the southern part of the Western Ghats in India, is lower than the hilly terrain to its north and south. The exact reasons for the formation of this gap are not clear. It result in the neighbouring regions of Tamil Nadu getting more rainfall from the South West monsoon and the neighbouring regions of Kerala having higher summer temperatures. What can be inferred from this passage?

(a) The Palghat gap is caused by high temperatures in southern Tamil Nadu and Kerala.
(b) The regions in Tamil Nadu and Kerala that are near the Palghat Gap are low lying.
(c) The low terrain of the Palghat Gap has a significant impact on weather patterns in neighbouring parts of Tamil Nadu and Kerala.
(d) Higher summer temperatures result in higher rainfall near the Palghat Gap area.

[EE & CS, 2014, 2 Marks (Set-2)]

57. Geneticists say that are very close to confirming the genetic roots of psychiatric illnesses such as depression and schizophrenia, and consequently, that doctors will be able to eradicate these diseases through early identification and gene therapy.

On which of the following assumptions dose the statement above rely?

(a) Strategies are now available for eliminating psychiatric illness.
(b) Certain psychiatric illness have a genetic basis.
(c) All human disease can be traced back to genes and how they are expressed.
(d) In the future, genetics will become the only relevant field for identifying psychiatric illness.

[EE & CS, 2014, 2 Marks (Set-2)]

58. Choose the most appropriate phrase from the options given below to complete the following sentence.

India is a post colonial country because

(a) it was a former British colony
(b) Indian Information Technology profession has colonized the world
(c) India does not follow any colonial practices
(d) India has helped other countries gain freedom

[EE & CS, 2014, 1 Mark (Set-2)]

59. Who ________ was coming to see us this evening?

(a) you said
(b) did you say
(c) did you say that
(d) had you said

[EE & CS, 2014, 1 Mark (Set-2)]

60. Match the columns:

Column-1	**Column-2**
1. eradicate	P. misrepresent
2. distort	Q. soak completely
3. saturate	R. use
4. utilize	S. destroy utterly

(a) 1-S, 2-P, 3-Q, 4-R
(b) 1-P, 2-Q, 3-R, 4-S
(c) 1-Q, 2-R, 3-S, 4-P
(d) 1-S, 2-P, 3-R, 4-Q

[EE & CS, 2014, 1 Mark (Set-2)]

61. The old city of Koenigsberg, which had a German majority population before World War 2, is now called Kaliningrad. After the events of the war, Kaliningrad is now a Russian territory and has a predominantly Russian population. It is bordered by the Baltic Sea on the north and the countries of Poland to the South and West and Lithuania to the East respectively. Which of the statements below can be inferred from this passage.

(a) Kaliningrad was historically Russian in its ethnic make up

(b) Kaliningrad is a part of Russia despite it not being contiguous with the rest of Russia

(c) Koenigsberg was renamed Kaliningrad, as that was its original Russian name

(d) Poland and Lithuania are on the route from Kaliningrad to the rest of Russia

[EE & CS, 2014, 2 Marks (Set-2)]

62. The number of people diagnosed with dengue fever (contracted from the bite of a mosquito) in North India is twice the number diagnosed last year. Municipal authorities have concluded that measures to control the mosquito population have failed in this region. **Which one of the following statements, if true, does not contradict this conclusion?**

(a) A high proportion of the affected population has returned from neighbouring countries where dengue is prevalent.

(b) More cases of dengue are now reported because of an increase in the Municipal Office's administrative efficiency.

(c) Many more cases of dengue are being diagnosed this year since the introduction of a new effective diagnostic test.

(d) The number of people with malarial fever (also contracted from mosquito bites) has increased this year.

[EE & CS, 2014, 2 Marks (Set-2)]

63. While trying to collect (I) an envelope from under the table (II), Mr. X fell down (III) and was losing consciousness (IV).

Which one of the above underlined parts of the sentence is NOT appropriate?

(a) I (b) II

(c) III (d) IV

[EE & CS, 2014, 1 Mark (Set-3)]

64. If she ______ how to calibrate the instrument, she _____ done the experiment.

(a) knows, will have

(b) knew, had

(c) had known, could have

(d) should have known, would have

[EE & CS, 2014, 1 Mark (Set-3)]

65. Choose the word that is opposite in meaning to the world "coherent"

(a) sticky (b) well-connected

(c) rambling (d) friendly

[EE & CS, 2014, 1 Mark (Set-3)]

66. A dance programme is scheduled for 10.00 a.m. Some students are participating in the programme and they need to come an hour earlier than the start of the event. These students should be accompanied by a parent. Other students and parents should come in time for the programme. The instruction you think that is appropriate for this is

(a) students should come at 9.00 am and parents should come at 10.00 am.

(b) participating students should come at 9.00 am accompanied by a parent, and other parents and students should come by 10.00 am.

(c) students who are not participating should come by 10.00 am and they should not bring their parent. Participating students should come at 9.00 am.

(d) participating students should come before 9.00 am. Parents who accompany them should come at 9.00 am. All other should come at 10.00 am.

[EE & CS, 2014, 2 Marks (Set-3)]

67. By the beginning of the 20th century, several hypotheses were being proposed, suggesting a paradigm shift in our understanding of the universe. However, the clinching evidence was provided by experimental measurements of the position of a star which was directly behind our sun.

Which of the following inference(s) may be drawn from the above passage?

(i) Out understanding of the universe changes based on the positions of stars
(ii) Paradigm shifts usually occur at the beginning of centuries
(iii) Stars are important objects in the universe
(iv) Experimental evidence was important in confirming this paradigm shift
(a) (i), (ii) and (iv) (b) (iii) only
(c) (i) and (iv) (d) (iv) only

[EE & CS, 2014, 2 Marks (Set-3)]

68. A student is required to demonstrate a high level of comprehension of the subject, especially in the social sciences.
The word closest in meaning to comprehension is
(a) understanding (b) meaning
(c) concentration (d) stability

[CE, 2014, 1 Mark (Set-1)]

69. Choose the most appropriate word from the options given below to complete the following sentence.
One of his biggest ______ was his ability to forgive.
(a) vice (b) virtues
(c) choices (d) strength

[CE, 2014, 1 Mark (Set-1)]

70. Rajan was not happy that Sajan decided to do the project on his own. On observing his unhappiness, Sajan explained to Rajan that he preferred to work independently.
Which one of the statements below is logically valid and can be inferred from the above sentences?
(a) Rajan has decided to work only in a group.
(b) Rajan and Sajan were formed into a group against their wishes.
(c) Sajan had decided to give in to Rajan's request to work with him.
(d) Rajan had believed that Sajan and he would be working together.

[CE, 2014, 1 Mark (Set-1)]

71. Choose the most appropriate word from the options given below to complete the following sentence.
A person suffering from Alzheimer's disease __________ short-term memory loss.
(a) experienced
(b) has experienced
(c) is experiencing
(d) experiences

[CE & IN, 2014, 1 Mark (Set-2)]

72. Choose the most appropriate word from the options given below to complete the following sentence.
__________ is the key to their happiness; they are satisfied with what they have.
(a) Contentment (b) Ambition
(c) Perseverance (d) Hunger

[CE & IN, 2014, 1 Mark (Set-2)]

73. Which of the following options is the closest in meaning to the sentence below?
"As a woman, I have no country."
(a) Women have no country.
(b) Women are not citizens of any country.
(c) Women's solidarity knows no national boundaries.
(d) Women of all countries have equal legal rights.

[CE & IN, 2014, 1 Mark (Set-2)]

74. Moving into a world of big data will require us to change our thinking about the merits of exactitude. To apply the conventional mindset of measurement to the digital, connected world of the twenty-first century is to miss a crucial point. As mentioned earlier, the obsession with exactness is an artefact of the information-deprived analog era. When data was sparse, every data point was critical, and thus great care was taken to avoid letting any point bias the analysis. From "BIG DATA" Viktor Mayer-Schonberger and Kenneth Cukier
The main point of the paragraph is:
(a) The twenty-first century is a digital world
(b) Big data is obsessed with exactness
(c) Exactitude is not critical in dealing with big data
(d) Sparse data leads to a bias in the analysis

[CE & IN, 2014, 2 Marks (Set-2)]

75. Select the pair that best expresses a relationship similar to that expressed in the pair
Children: Pediatrician
(a) Adult: Orthopaedist
(b) Females: Gynaecologist
(c) Kidney: Nephrologist
(d) Skin: Dermatologist
[CE (Set-1) & CS (Set-3), 2015, 1 Mark]

76. Extreme focus on syllabus and studying for tests has become such a dominant concern of Indian students that they close their minds to anything _______ to the requirements of the exam.
(a) related (b) extraneous
(c) outside (d) useful
[CE (Set-1) & CS (Set-3), 2015, 1 Mark]

77. The Tamil version of _______ John Abraham-starrer *Madras Cafe* _______ cleared by the Censor Board with no cuts last week, but the film's distributors _______ no takers among the exhibitors for a release in Tamil Nadu _______ this Friday.
(a) Mr., was, found, on
(b) a, was, found, at
(c) the, was, found, on
(d) a, being, find at
[CE (Set-1) & CS (Set-3), 2015, 1 Mark]

78. Most experts feel that in spite of possessing all the technical skills required to be a batsman of the highest order, he is unlikely to be so due to lack of requisite temperament. He was guilty of throwing away his wicket several times after working hard to lay a strong foundation. His critics pointed out that until he addressed this problem, success at the highest level will continue to elude him.
Which of the statement(s) below is/are logically valid and can be inferred from the above passage?
(i) He was already a successful batsman at the highest level.
(ii) He has to improve his temperament in order to become a great batsman.
(iii) He failed to make many of his good starts count.
(iv) Improving his technical skills will guarantee success.
(a) (iii) and (iv) (b) (ii) and (iii)
(c) (i), (ii) and (iii) (d) (ii) only
[CE (Set-1) & CS (Set-3), 2015, 2 Marks]

79. Alexander turned his attention towards India, since he had conquered Persia.
Which one of the statements below is logically valid and can be inferred from the above sentence?
(a) Alexander would not have turned his attention towards India had he not conquered Persia.
(b) Alexander was not ready to rest on his laurels, and wanted to march to India.
(c) Alexander was completely in control of his army and could command it to move towards India.
(d) Since Alexander's kingdom extended to Indian borders after the conquest of Persia, he was keen to move further.
[CE (Set-1) & CS (Set-3), 2015, 2 Marks]

80. Choose the statement where underlined word is used correctly
(a) The minister <u>insured</u> the victims that everything would be all right.
(b) He <u>ensured</u> that the company will not have to bear any loss.
(c) The actor got himself <u>ensured</u> against any accident.
(d) The teacher <u>insured</u> students of good results.
[CE 2015, 1 Mark (Set-2)]

81. Choose the most appropriate word from the options given below to complete the following sentence
The official answered _______ that the complaints of the citizen would be looked into.
(a) respectably (b) respectfully
(c) reputably (d) respectively
[CE 2015, 1 Mark (Set-2)]

82. The word similar in meaning to 'dreary' is
(a) cheerful (b) dreamy
(c) hard (d) dismal
[CE 2015, 2 Marks (Set-2)]

83. We _______ our friend's birthday and we _______ how to make it up to him
(a) completely forgot - - - don't just know
(b) forgot completely - - - don't just know
(c) completely forgot - - - just don't know
(d) forgot completely - - - just don't know
[CS (Set-1) & EE (Set-2), 2015, 1 Mark]

84. A generic term that includes various items of clothing such as a skirt, a pair of trousers and a shirt is
(a) fabric (b) textile
(c) fibre (d) apparel
[CS (Set-1) & EE (Set-2), 2015, 1 Mark]

85. Choose the statement where underlined word is used correctly.
(a) The industrialist had a personnel jet.
(b) I write my experience in my personnel diary.
(c) All personnel are being given the day off.
(d) Being religious is a personnel aspect.
[CS (Set-1) & EE (Set-2), 2015, 1 Mark]

86. Out of the following four sentences, select the most suitable sentence with respect to grammar and usage.
(a) Since the report lacked needed information, it was of no use to them.
(b) The report was useless to them because there were no needed information in it.
(c) Since the report did not contain the needed information, it was not real useful to them.
(d) Since the report lacked needed information, it would not had been useful to them.
[CS (Set-1) & EE (Set-2), 2015, 2 Marks]

87. Which one of the following combinations is incorrect?
(a) Acquiescence - Submission
(b) Wheedle - Roundabout
(c) Flippancy - Lightness
(d) Profligate - Extravagant
[CS (Set-2) & EE (Set-1), 2015, 1 Mark]

88. Didn't you buy _______ when you went shopping?
(a) any paper (b) much paper
(c) no paper (d) a few paper
[CS (Set-2) & EE (Set-1), 2015, 1 Mark]

89. Which of the following options is the closest in meaning to the sentence below?
She enjoyed herself immensely at the party.
(a) She had a terrible time at the party
(b) She had a horrible time at the party
(c) She had a terrific time at the party
(d) She had a terrifying time at the party
[CS (Set-2) & EE (Set-1), 2015, 1 Mark]

90. Select the alternative meaning of the underlined part of the sentence.
The chain snatchers took to their heels when the police party arrived.
(a) took shelter in a thick jungle
(b) open indiscriminate fire
(c) took to flight
(d) unconditionally surrendered
[CS (Set-2) & EE (Set-1), 2015, 2 Marks]

91. The given statement is followed by some courses of action. Assuming the statement to be true, decide the correct option.
Statement: There has been a significant drop in the water level in the lakes supplying water to the city.
Course of action:
(I) The water supply authority should impose a partial cut in supply to tackle the situation.
(II) The government should appeal to the all the residents through mass media for minimal use of water.
(III) The government should ban the water supply in lower areas.
(a) Statements I and II follow
(b) Statements I and III follow
(c) Statements II and III follow
(d) All statements follow
[CS (Set-2) & EE (Set-1), 2015, 2 Marks]

92. Choose the appropriate word/phrase, out of the four options given below, to complete the following sentence:
Frogs _______________
(a) croak (b) roar
(c) hiss (d) patter
[EC 2015, 1 Mark (Set-1)]

93. Choose the word most similar meaning to the given word:
Educe
(a) Exert (b) Educate
(c) Extract (d) Extend
[EC 2015, 1 Mark (Set-1)]

94. Choose the most appropriate word from the options given below to complete the following sentence.
The principal presented the chief guest with a __________, as token of appreciation.
(a) momento (b) memento
(c) momentum (d) moment
[EC 2015, 1 Mark (Set-1)]

95. The following question presents a sentence, part of which is underlined. Beneath the sentence you find four ways of phrasing the underlined part. Following the requirements of the standard written English, select the answer that produces the most effective sentence.

Tuberculosis, together with its effects, <u>ranks one of the leading causes of death</u> in India.

(a) ranks as one of the leading causes of death
(b) rank as one of the leading causes of death
(c) has the rank of one of the leading causes of death
(d) are one of the leading causes of death

[EC 2015, 2 Marks (Set-1)]

96. Humpty Dumpty sits on a wall every day while having lunch. The wall sometimes breaks. A person sitting on the wall falls if the wall breaks. Which one of the statements below is logically valid and can be inferred from the above sentences?

(a) Humpty Dumpty always falls white having lunch
(b) Humpty Dumpty does not fall sometimes while having lunch
(c) Humpty Dumpty never falls during dinner
(d) When Humpty Dumpty does not sit on the wall, the wall does not break

[EC 2015, 2 Marks (Set-1)]

97. Read the following paragraph and choose the correct statement.

Climate change has reduced human security and threatened human well being. An ignored reality of human progress is that human security largely depends upon environmental security. But on the contrary, human progress seems contradictory to environmental security. To keep up both at the required level is a challenge to be addressed by one and all. One of the ways to curb the climate change may be suitable scientific innovations, while the other may be the Gandhian perspective on small scale progress with focus on sustainability.

(a) Human progress and security are positively associated with environmental security.
(b) Human progress is contradictory to environmental security.
(c) Human security is contradictory to environmental security.
(d) Human progress depends upon environmental security.

[EC 2015, 2 Marks (Set-1)]

98. Choose the word most similar in meaning to the given word:

Awkward

(a) Inept (b) Graceful
(c) Suitable (d) Dreadful

[EC (Set-2) & ME (Set-3), 2015, 1 Mark]

99. Choose the appropriate word/phase, out of the four options given below, to complete the following sentence:

Dhoni, as well as the other team members of the Indian team _____ present on the occasion.

(a) were (b) was
(c) has (d) have

[EC (Set-2) & ME (Set-3), 2015, 1 Mark]

100. What is the adverb for the given word below? Misogynous

(a) Misogynousness (b) Misogynity
(c) Misogynously (d) Misogynous

[EC (Set-2) & ME (Set-3), 2015, 1 Mark]

101. Lamenting the gradual sidelining of the arts in school curricula, a group of prominent artists wrote to the Chief Minister last year, asking him to allocate more funds to support arts education in schools. However, no such increase has been announced in this year's Budget. The artists expressed their deep anguish at their request not being approved, but many of them remain optimistic about finding in the future.

Which of the statement(s) below is/are logically valid and can be inferred from the above statements?

(i) The artists expected funding for the arts to increase this year.
(ii) The Chief Minister was receptive to the idea of increasing funding for the arts.
(iii) The Chief Minister is a prominent artists.
(iv) Schools are giving less importance to arts education nowadays.

(a) (iii) and (iv) (b) (i) and (iv)
(c) (i), (ii) and (iv) (d) (i) and (iii)

[EC (Set-2) & ME (Set-3), 2015, 2 Marks]

102. Choose the correct verb to fill in the blank below:
Let us ______.
(a) introvert (b) alternate
(c) atheist (d) altruist
[EC (Set-3) & ME (Set-2), 2015, 1 Mark]

103. Choose the most appropriate word from the options given below to complete the following sentence.
If the athlete had wanted to come first in the race, he ______ several hours every day.
(a) should practice
(b) should have practised
(c) practised
(d) should be practising
[EC (Set-3) & ME (Set-2), 2015, 1 Mark]

104. Choose the most suitable one word substitute for the following expression:
Connection of a road or way
(a) Pernicious (b) Viaticum
(c) Clandestine (d) Ravenous
[EC (Set-3) & ME (Set-2), 2015, 1 Mark]

105. Ram and Shyam shared a secret and promised to each other that it would remain between them. Ram express himself in one of the following ways as given in the choices below. Identify the correct way as per standard English.
(a) It would remain between you and me.
(b) It would remain between I and you.
(c) It would remain between you and I.
(d) It would remain with me.
[EC (Set-3) & ME (Set-2), 2015, 2 Marks]

106. In the following question, the first and the last sentence of the passage are in order and numbered 1 and 6. The rest of the passage is split into 4 parts and numbered as 2, 3, 4 and 5. These 4 parts are not arranged in proper order. Read the sentences and arrange them in a logical sequence to make a passage and choose the correct sequence from the given options.
1. On Diwali, the family rises early in the morning.
2. The whole family, including the young and the old enjoy doing this.
3. Children let off fireworks later in the night with their friends.
4. At sunset, the lamps are lit and the family performs various rituals.
5. Father, mother and children visit relatives and exchange gifts and sweets.
6. Houses looks so pretty with lighted lamps all around.
(a) 2, 5, 3, 4 (b) 5, 2, 4, 3
(c) 3, 5, 4, 2 (d) 4, 5, 2, 3
[EC (Set-3) & ME (Set-2), 2015, 2 Marks]

107. Choose the appropriate word/phrase, out of the four options given below, to complete the following sentence:
Apparent lifelessness ______ dormant life.
(a) harbours (b) leads to
(c) supports (d) affects
[ME (Set-1), IN & PI 2015, 1 Mark]

108. Choose the statement where underlined word is used correctly:
(a) When the teacher eludes to different authors, he is being <u>elusive</u>.
(b) When the thief keeps eluding the police, he is being <u>elusive</u>.
(c) Matters the are difficult to understand, identify or remember are <u>allusive</u>.
(d) Mirages can be <u>allusive</u>, but a better way to express them is illusory.
[ME (Set-1), IN & PI 2015, 1 Mark]

109. Fill in the blank with the correct idiom/phrase.
That boy from the two was a ______ in the sleepy village.
(a) dog out of herd
(b) sheep from the heap
(c) fish out of water
(d) bird from the flock
[ME (Set-1), IN & PI 2015, 1 Mark]

110. Select the appropriate option in place of underlined part of the sentence:
<u>Increased productivity necessary</u> reflects greater efforts made by the employees.
(a) Increase in productivity necessary
(b) Increase productivity is necessary
(c) Increase in productivity necessarily
(d) No improvement required
[ME (Set-1), IN & PI 2015, 2 Marks]

111. Which of the following is **CORRECT** with respect to grammar and usage?
Mount Everest is________.
(a) the highest peak in the world
(b) highest peak in the world
(c) one of highest peak in the world
(d) one of the highest peak in the world
[EC & ME 2016, 1 Mark (Set-1)]

112. The policeman asked the victim of a theft, "What did you________?"
(a) loose (b) lose
(c) loss (d) louse
[EC & ME 2016, 1 Mark (Set-1)]

113. Despite the new medicine's_____ in treating diabetes, it is not_______ widely.
(a) effectiveness---prescribed
(b) availability---used
(c) prescription---available
(d) acceptance---proscribed
[EC & ME 2016, 1 Mark (Set-1)]

114. In a world filled with uncertainty, he was glad to have many good friends. He had always assisted them in times of need and was confident that they would reciprocate. However, the events of the last week proved him wrong.
Which of the following inference(s) is/are logically valid and can be inferred from the above passage?
(i) His friends were always asking him to help them.
(ii) He felt that when in need of help, his friends would let him down.
(iii) He was sure that his friends would help him when in need.
(iv) His friends did not help him last week.
(a) (i) and (ii) (b) (iii) and (iv)
(c) (iii) only (d) (iv) only
[EC & ME 2016, 2 Marks (Set-1)]

115. Based on the given statements, select the appropriate option with respect to grammar and usage.
Statements
(i) The height of Mr. X is 6 feet.
(ii) The height of Mr. Y is 5 feet.
(a) Mr. X is longer than Mr. Y.
(b) Mr. X is more elongated than Mr. Y.
(c) Mr. X is taller than Mr. Y.
(d) Mr. X is lengthier than Mr. Y.
[EC (Set-2) & ME (Set-3) 2016, 1 Mark]

116. The students __________ the teacher on teachers' day for twenty years of dedicated teaching.
(a) facilitated (b) felicitated
(c) fantasized (d) facilitated
[EC (Set-2) & ME (Set-3) 2016, 1 Mark]

117. After India's cricket world cup victory in 1985, Shrotria who was playing both tennis and cricket till then, decided to concentrate only on cricket. <u>And the rest is history.</u>
What does the underlined phrase mean in this context?
(a) history will rest in peace
(b) rest is recorded in history books
(c) rest is well known
(d) rest is archaic
[EC (Set-2) & ME (Set-3) 2016, 1 Mark]

118. Social science disciplines were in an amorphous form until the colonial period when they were institutionalized. In varying degree, they were intended to further the colonial interest. In the time of globalization and the economic rise of postcolonial countries like India, conventional ways of knowledge production have become obsolete.
Which of the following can be logically inferred from the above statements?
(i) Social science disciplines have become obsolete
(ii) Social science disciplines had a pre-colonial origin
(iii) Social science disciplines always promote colonialism.
(iv) Social science must maintain disciplinary boundaries.
(a) (ii) only
(b) (i) and (iii) only
(c) (ii) and (iv) only
(d) (iii) and (iv) only
[EC (Set-2) & ME (Set-3) 2016, 2 Marks]

119. *M* has a son *Q* and a daughter *R*. He has no other children. *E* is the mother of *P* and daughter-in-law of *M*. How is *P* related to *M*?
(a) *P* is the son-in-law of *M*
(b) *P* is the grandchild of *M*
(c) *P* is the daughter-in law of *M*
(d) *P* is the grandfather of *M*
[EC (Set-3) & IN 2016, 1 Mark]

120. An apple costs ₹ 10. An onion costs ₹ 8.
Select the most suitable sentence with respect to grammar and usage.
(a) The price of an apple is greater than an onion.
(b) The price of an apple is more than onion.
(c) The price of an apple is greater than that of an onion.
(d) Apples are more costlier than onions.
[EC (Set-3) & IN 2016, 1 Mark]

121. The Buddha said, "Holding on to anger is like grasping a hot coal with the intent of throwing it at someone else; you are the one who gets burnt."
Select the word below which is closest in meaning to the word underlined above.
(a) burning (b) igniting
(c) clutching (d) flinging
[EC (Set-3) & IN 2016, 1 Mark]

122. The overwhelming number of people infected with rabies in India has been flagged by the World Health Organization as a source of concern. It is estimated that inoculating 70% of pets and stray dogs against rabies can lead to a significant reduction in the number of people infected with rabies.
Which of the following can be logically inferred from the above sentences?
(a) The number of people in India infected with rabies is high.
(b) The number of people in other parts of the world who are infected with rabies is low
(c) Rabies can be eradicated in India by vaccinating 70% of stray dogs
(d) Stray dogs are the main source of rabies worldwide
[EC (Set-3) & IN 2016, 2 Marks]

123. Out of the following four sentences, select the most suitable sentence with respect to grammar and usage.
(a) I will not leave the place until the minister does not meet me.
(b) I will not leave the place until the minister doesn't meet me.
(c) I will not leave the place until the minister meet me.
(d) I will not leave the place until the minister meets me.
[CE & CS (Set-1), 2016, 1 Mark]

124. A rewording of something written or spoken is a __________.
(a) paraphrase (b) paradox
(c) paradigm (d) paraffin
[CE & CS (Set-1), 2016, 1 Mark]

125. Archimedes said, "Give me a lever long enough and a fulcrum on which to place it, and I will move the world." The sentence above is an example of a __________ statement.
(a) figurative
(b) collateral
(c) literal
(d) figurine
[CE & CS (Set-1), 2016, 1 Mark]

126. Indian currency notes show the denomination indicated in at least seventeen languages. If this is not an indication of the nation's diversity, nothing else is. Which of the following can be logically inferred from the above sentences?
(a) India is a country of exactly seventeen languages.
(b) Linguistic pluralism is the only indicator of a nation's diversity.
(c) Indian currency notes have sufficient space for all the Indian languages.
(d) Linguistic pluralism is strong evidence of India's diversity.
[CE & CS (Set-1), 2016, 2 Marks]

127. The man who is now Municipal Commissioner worked as __________.
(a) the security guard at a university
(b) a security guard at the university
(c) a security guard at university
(d) the security guard at the university
[EE (Set-1) & CS (Set-2), 2016, 1 Mark]

128. Nobody knows how the Indian cricket team is going to cope with the difficult and seamer-friendly wickets in Australia.
Choose the option which is closest in meaning to the underlined phrase in the above sentence.
(a) put up with (b) put in with
(c) put down to (d) put up against
[EE (Set-1) & CS (Set-2), 2016, 1 Mark]

129. Find the odd one in the following group of words.
mock, deride, praise, jeer
(a) mock (b) deride
(c) praise (d) jeer
[EE (Set-1) & CS (Set-2), 2016, 1 Mark]

130. Computers were invented for performing only high-end useful computations. However, it is no understatement that they have taken over our world today. The internet, for example, is ubiquitous. Many believe that the internet itself is an unintended consequence of the original invention. With the advent of mobile computing on our phones, a whole new dimension is now enabled. One is left wondering if all these developments are good or, more importantly, required.
Which of the statement(s) below is/are logically valid and can be inferred from the above paragraph?
(i) The author believes that computers are not good for us.
(ii) Mobile computers and the internet are both intended inventions
(a) (i) only
(b) (ii) only
(c) both (i) and (ii)
(d) neither (i) nor (ii)
[EE (Set-1) & CS (Set-2), 2016, 2 Marks]

131. All hill-stations have a lake. Ooty has two lakes. Which of the statement(s) below is/are logically valid and can be inferred from the above sentences?
(i) Ooty is not a hill-station.
(ii) No hill-station can have more than one lake.
(a) (i) only
(b) (ii) only
(c) both (i) and (ii)
(d) neither (i) nor (ii)
[EE (Set-1) & CS (Set-2), 2016, 2 Marks]

132. If I were you, I _____ that laptop. It's much too expensive.
(a) won't buy (b) shan't buy
(c) wouldn't buy (d) would by
[CE (Set-2), 2016, 1 Mark]

133. He turned a deaf ear to my request.
What does the underlined phrasal verb mean?
(a) ignored (b) appreciated
(c) twisted (d) returned
[CE (Set-2), 2016, 1 Mark]

134. Choose the most appropriate set of words from the options given below to complete the following sentence.
_____ _____ is a will, _____ is a way.
(a) wear, there, their
(b) were, their, there
(c) where, there, there
(d) were, their, their
[CE (Set-2), 2016, 1 Mark]

135. Today we consider, Ashoka as a great ruler because of the copious evidence he left behind in the form of stone carved edicts. Historians tend to correlate greatness of a king at his time with the availability of evidence today.
Which of the following can be logically inferred from the above sentences?
(a) Emperors who do not leave significant sculpted evidence are completely forgotten.
(b) Ashoka produced stone carved edicts to ensure that later historians will respect him.
(c) Statues of kings are a reminder of their greatness.
(d) A king's greatness, as we know him today, is interpreted by historians.
[CE (Set-2), 2016, 2 Marks]

136. The chairman requested the aggrieved shareholders to _________ him.
(a) bare with (b) bore with
(c) bear with (d) bare
[EE (Set-2), 2016, 1 Mark]

137. Identify the correct spelling out of the given options:
(a) Managable (b) Manageable
(c) Mangaeble (d) Managible
[EE (Set-2), 2016, 1 Mark]

138. R2D2 is a robot. R2D2 can repair aeroplanes. No other robot can repair aeroplanes. Which of the following can be logically inferred from the above statements?

(a) R2D2 is a robot which can only repair aeroplanes.
(b) R2D2 is the only robot which can repair aeroplanes.
(c) R2D2 is a robot which can repair only aeroplanes.
(d) Only R2D2 is a robot.

[EE (Set-2), 2016, 1 Mark]

139. A poll of students appearing for masters in engineering indicated that 60 % of the students believed that mechanical engineering is a profession unsuitable for women. A research study on women with masters or higher degrees in mechanical engineering found that 99 % of such women were successful in their professions.

Which of the following can be logically inferred from the above paragraph?

(a) Many students have misconceptions regarding various engineering disciplines.
(b) Men with advanced degrees in mechanical engineering believe women are well suited to be mechanical engineers.
(c) Mechanical engineering is a profession well suited for women with masters or higher degrees in mechanical engineering.
(d) The number of women pursuing higher degrees in mechanical engineering is small.

[EE (Set-2), 2016, 2 Marks]

140. Sourya committee had proposed the establishment of Sourya Institutes of Technology (SITs) in line with Indian Institutes of Technology (IITs) to cater to the technological and industrial needs of a developing country.

Which of the following can be logically inferred from the above sentence?

Based on the proposal,

(i) In the initial years, SIT students will get degrees from IIT.
(ii) SITs will have a distinct national objective.
(iii) SIT like institutions can only be established in consultation with IIT.
(iv) SITs will serve technological needs of a developing country.

(a) (iii) and (iv) only.
(b) (i) and (iv) only.
(c) (ii) and (iv) only
(d) (ii) and (iii) only

[EE (Set-2), 2016, 2 Marks]

141. The volume of a sphere of diameter 1 unit is ________ than the volume of a cube of side 1 unit.

(a) least
(b) less
(c) lesser
(d) low

[ME (Set-2), 2016, 1 Mark]

142. The unruly crowd demanded that the accused be ___________ without trial.

(a) hanged
(b) hanging
(c) hankering
(d) hung

[ME (Set-2), 2016, 1 Mark]

143. Choose the statement(s) where the underlined word is used correctly:

(i) A <u>prone</u> is a dried plum.
(ii) He was lying <u>prone</u> on the floor.
(iii) People who eat a lot of fat are <u>prone</u> to heart disease.

(a) (i) and (iii) only
(b) (iii) only
(c) (i) and (ii) only
(d) (ii) and (iii) only

[ME (Set-2), 2016, 1 Mark]

144. A smart city integrates all modes of transport, uses clean energy and promotes sustainable use of resources. It also uses technology to ensure safety and security of the city, something which critics argue, will lead to a surveillance state. Which of the following can be logically inferred from the above paragraph?

(i) All smart cities encourage the formation of surveillance states.
(ii) Surveillance is an integral part of a smart city.
(iii) Sustainability and surveillance go hand in hand in a smart city.
(iv) There is a perception that smart cities promote surveillance.

(a) (i) and (iv) only
(b) (ii) and (iii) only
(c) (iv) only
(d) (i) only

[ME (Set-2), 2016, 2 Marks]

145. The bacteria in milk are destroyed when it ___________ heated to 80 degree Celsius.

(a) would be
(b) will be
(c) is
(d) was

[CE 2017, 1 Mark (Set-1)]

146. ________ with someone else's email account is now a very serious offense.
(a) Involving (b) Assisting
(c) Tampering (d) Incubating
[CE, 2017, 1 Mark (Set-1)]

147. The old concert hall was demolished because of fears that the foundation would be affected by the construction of the new metro line in the area. Modern technology for underground metro construction tried to mitigate the impact of pressurized air pockets created by the excavation of large amounts of soil. But even with these safeguards, it was feared that the soil below the concert hall would not be stable. From this, one can infer that
(a) the foundations of old buildings create pressurized air pockets underground, which are different to handle during metro construction.
(b) metro construction has to be done carefully considering its impact on the foundations of existing buildings.
(c) old buildings in an area form an impossible hurdle to metro construction in that area.
(d) pressurized air can be used to excavate large amount of soil from underground areas.
[CE, 2017, 2 Marks (Set-1)]

148. The event would have been successful if you ________ able to come.
(a) are (b) had been
(c) have been (d) would have been
[CE (Set-2) & IN, 2017, 1 Mark]

149. There was no doubt that their work was thorough. Which of the words below is closest in meaning to the underlined word above?
(a) pretty (b) complete
(c) sloppy (d) haphazard
[CE (Set-2) & IN, 2017, 1 Mark]

150. Bhaichung was observing the pattern of people entering and leaving a car service centre. There was a single window where customers were being served. He saw that people inevitably came out of the centre in the order that they went in. However, the time they spent inside seemed to vary a lot: Some people came out in a matter of minutes while for others it took much longer.
From this, what can one conclude?
(a) The centre operates on a first-come-first-served basis, but with variable service times, depending on specific customer needs.
(b) Customers were served in an arbitrary order, since they took varying amounts of time for service completion in the centre.
(c) Since some people came out within a few minutes of entering the centre, the system is likely to operate on a last-come-first-served basis.
(d) Entering the centre early ensured that one would have shorter service times and most people attempted to do this.
[CE (Set-2) & IN, 2017, 2 Marks]

151. After Rajendra Chola returned from his voyage to Indonesia, he ________ to visit the temple in Thanjavur.
(a) was wishing (b) is wishing
(c) wished (d) had wished
[CS & EE 2017, 1 Mark (Set-1)]

152. Research in the workplace reveal that people work for many reasons ________.
(a) money beside (b) beside money
(c) money besides (d) besides money
[CS & EE 2017, 1 Mark (Set-1)]

153. "The colonial past is such that anything inadequately or improperly nationalist is just not history."
Which of the following statements best reflects the author's opinion?
(a) Nationalists are highly imaginative
(b) History is viewed through the filter of nationalism
(c) Our colonial past never happened
(d) Nationalism has to be both adequately and properly imagined.
[CS & EE 2017, 2 Marks (Set-1)]

154. Saturn is ________ to be seen on a clear night with the naked eye.
(a) enough bright (b) bright enough
(c) as enough bright (d) bright as enough
[CS & EE 2017, 1 Mark (Set-2)]

155. Choose the option with words that are not synonyms.
(a) aversion, dislike
(b) luminous, radiant
(c) plunder, loot
(d) yielding, resistant
[CS & EE 2017, 1 Mark (Set-2)]

156. "We lived in a culture that denied any merit to literary works, considering them important only when they were handmaidens to something seemingly more urgent-namely ideology. This was a country where all gestures, even the most private, were interpreted in political terms,".
The author's belief that ideology is not as important as literature is revealed by the word:
(a) 'culture' (b) 'seemingly'
(c) 'urgent' (d) 'political'
[CS & EE 2017, 2 Marks (Set-2)]

157. I __________ made arrangements had I __________ informed earlier.
(a) could have, been (b) would have, being
(c) had, have (d) had been, been
[EC 2017, 1 Mark (Set-1)]

158. She has a sharp tongue and it can occasionally turn __________.
(a) hurtful (b) left
(c) methodical (d) vital
[EC 2017, 1 Mark (Set-1)]

159. "If you are looking for a history of India, or for an account of the rise and fall of the British Raj, or for the reason of the cleaving of the subcontinent into two mutually antagonistic parts and the effects this mutilation will have in the respective sections, and ultimately on Asia, you will not find it in these pages; for though I have spent a lifetime in the country, I lived too near the seat of events, and was too intimately associated with the actors, to get the perspective needed for the impartial recording of these matters".
Here, the word 'antagonistic' is closest in meaning to
(a) impartial (b) argumentative
(c) separated (d) hostile
[EC 2017, 2 Marks (Set-1)]

160. It is __________ to read this year's textbook __________ the last year's.
(a) easier, than (b) most easy, than
(c) easier, from (d) easiest, from
[EC 2017, 1 Mark (Set-2)]

161. The ninth and the tenth of this month are Monday and Tuesday _____.
(a) figuratively (b) retrospectively
(c) respectively (d) rightfully
[EC 2017, 1 Mark (Set-2)]

162. "If you are looking for a history of India, or for an account of the rise and fall of the British Raj or for the reason of the cleaving of the subcontinent into two mutually antagonistic parts and the effects this mutilation will have in the respective sections, and ultimately on Asia, you will not find it in these pages: for though I have spent a lifetime in the country. I lived too near the seat of events, and was too intimately associated with the actors, to get the perspective needed for the impartial recording of these matters."
Which of the following statements best reflects the author's opinion?
(a) An intimate association does not allow for the necessary perspective.
(b) Matters are recorded with an impartial perspective.
(c) An intimate association offers an impartial perspective.
(d) Actors are typically associated with the impartial recording of matters.
[EC 2017, 2 Marks (Set-2)]

163. He was one of my best ________ and I felt his loss ________.
(a) friend, keenly
(b) friends, keen
(c) friend, keener
(d) friends, keenly
[ME 2017, 1 Mark (Set-1]

164. As the two speakers became increasingly agitated, the debate became _______.
(a) lukewarm (b) poetic
(c) forgiving (d) heated
[ME 2017, 1 Mark (Set-1]

165. Two very famous sportsmen Mark and Steve happened to be brothers, and played for country K. Mark teased James, and opponent from country E. There is no way you are good enough to play for your country. James replied, 'May be not, but at least I am the best player in my own family'.
Which one of the following can be inferred from this conversation.
(a) Mark was known to play better than James
(b) Steve was known to play better than Mark
(c) James and Steve were good friends
(d) James played better than Steve
[ME 2017, 2 Marks (Set-1]

166. The ways in which this game can be played ______ potentially infinite.
(a) is (b) is being
(c) are (d) are being
[ME 2017, 1 Mark (Set-2]

167. If you choose plan P, you will have to ______ plan Q, as these two are mutually _____.
(a) forgo, exclusive
(b) forget, inclusive
(c) accept, exhaustive
(d) adopt, intrusive
[ME 2017, 1 Mark (Set-2]

168. "The driver applied the ______ as soon as she approached the hotel where she wanted to take a ________."
The words that best fill the blanks in the above sentence are
(a) brake, break (b) break, break
(c) brake, brake (d) break, brake
[CE, 2018, 1 Mark (Set-1)]

169. "It is no surprise that every society has had codes of behaviour; however, the nature of these codes is often ________."
The word that best fills the blank in the above sentence is
(a) unpredictable (b) simple
(c) expected (d) strict
[CE, 2018, 1 Mark (Set-1)]

170. "His face ________ with joy when the solution of the puzzle was ________ to him." The words that best fill the blanks in the above sentence are
(a) shone, shown (b) shone, shone
(c) shown, shone (d) shown, shown
[CE, 2018, 1 Mark (Set-2)]

171. "Although it does contain some pioneering ideas, one would hardly characterize the work as ________."
The word that best fills the blank in the above sentence is
(a) innovative (b) simple
(c) dull (d) boring
[CE, 2018, 1 Mark (Set-2)]

172. "From where are they bringing their books? ______ bringing ______ books from ______."
The words that best fill the blanks in the above sentence are
(a) Their, they're, there
(b) They're, their, there
(c) There, their, they're
(d) They're, there, there
[CS, 2018, 1 Mark]

173. "A ______ investigation can sometimes yield new facts, but typically organized ones are more successful."
The word that best fills the blank in the above sentence is
(a) meandering (b) timely
(c) consistent (d) systematic
[CS, 2018, 1 Mark]

174. "By giving him the last _____ of the cake, you will ensure lasting _____ in our house today."
The words that best fill the blanks in the above sentence are
(a) peas, piece(b) piece, peace
(c) peace, piece(d) peace, peas
[EC, 2018, 1 Mark]

175. "Even though there is a vast scope for its ________, tourism has remained a/an _____ area."
The words that best fill the blanks in the above sentence are
(a) improvement, neglected
(b) rejection, approved
(c) fame, glum
(d) interest, disinterested
[EC, 2018, 1 Mark]

176. "Since you have gone off the _____, the ______ sand is likely to damage the car". The words that best fill the blanks in the above sentence are

(a) course, coarse (b) course, course
(c) coarse, course (d) coarse, coarse

[EE, 2018, 1 Mark]

177. "A common misconception among writers is that sentence structure mirrors thought; the more ______ the structure, the more complicated the ideas".

The word that best fills the blank in the above sentence is

(a) detailed (b) simple
(c) clear (d) convoluted

[EE, 2018, 1 Mark]

178. "In spite of being warned repeatedly, he failed to correct his __________ behaviour." The word that best fills that blank in the above sentence is

(a) rational (b) reasonable
(c) errant (d) good

[IN, 2018, 1 Mark]

179. "When she fell down the __________. She received many __________ but little help." The words that best fill the blanks in the above sentence are

(a) stairs, stares (b) stairs, stairs
(c) stares, stairs (d) stares, stares

[IN, 2018, 1 Mark]

180. "Her ______ should not be confused with miserliness; she is ever willing to assist those in need."

The word that best fills the blank in the above sentence is:

(a) cleanliness (b) punctuality
(c) frugality (d) greatness

[ME, 2018, 1 Mark (Set-1)]

181. Going by the _____that many hands make light work, the school______involved all the students in the task."

(a) principle, principal
(b) principal, principle
(c) principle, principle
(d) principal, principal

[ME, 2018, 1 Mark (Set-1)]

182. "The dress _______ her so well that they all immediately ______ her on her appearance."

The words that best fill the blanks in the above sentence are

(a) complemented, complemented
(b) complimented, complemented
(c) complimented, complimented
(d) complemented, complimented

[ME, 2018, 1 Mark (Set-2)]

183. "The judge's standing in the legal community, though shaken by false allegations of wrongdoing, remained ________ ."

The word that best fill the blanks in the above sentence is

(a) undiminished (b) damaged
(c) illegal (d) uncertain

[ME, 2018, 1 Mark (Set-2)]

184. They have come a long way in ________ trust among the users.

(a) creating (b) created
(c) creation (d) create

[CE, 2019 : 1 Mark, Set-I]

185. The lecture was attended by quite _______ students, so the hall was not very _____.

(a) few, quite (b) a few, quite
(c) few, quiet (d) a few, quiet

[CE, 2019 : 1 Mark, Set-I]

186. The CEO's decision to quit was as shocking to Board as it was to __________.

(a) myself (b) I
(c) me (d) my

[CE, 2019 : 1 Mark, Set-I]

187. Daytime temperatures in Delhi can______40°C.

(a) get (b) peak
(c) reach (d) stand

[CE, 2019 : 1 Mark, Set-II]

188. Hima Das was_____only Indian athlete to win_______gold for India.

(a) the, a (b) an, a
(c) an, the (d) the, many

[CE, 2019 : 1 Mark, Set-II]

189. The growth rate of ABC Motors in 2017 was the same__________XYZ Motors in 2016.

(a) as that off (b) as those of
(c) as off (d) as that of

[CE, 2019 : 1 Mark, Set-II]

190. "The increasing interest in tribal characters might be a mere coincidence, but the timing is of interest. None of this, though, is to say that the tribal hero has arrived in Hindi cinema, or that the new crop of characters represents the acceptance of the tribal character in the industry. The films and characters are too few to be described as a pattern."

What does the word 'arrived' mean in the paragraph above?

(a) came to a conclusion
(b) reached a terminus
(c) went to a place
(d) attained a status

[CE, 2019 : 2 Marks, Set-I]

191. The new cotton technology, Bollgard-II, with herbicide tolerant traits has developed into a thriving business in India. However, the commercial use of this technology is not legal in India. Notwithstanding that, reports indicate that the herbicide tolerant Bt cotton had been purchased by farmers at an average of Rs 200 more than the control price of ordinary cotton, and planted in 15% of the cotton growing area in the 2017 Kharif season.

Which one of the following statements can be inferred from the given passage?

(a) Farmers want to access the new technology for experimental purposes.
(b) Farmers want to access the new technology if India benefits from it.
(c) Farmers want to access the new technology even if it is not legal.
(d) Farmers want to access the new technology by paying high price.

[CE, 2019 : 2 Marks, Set-I]

192. "Popular Hindi fiction, despite - or perhaps because of - its wide reach, often does not appear in our cinema. As ideals that viewers are meant to look up to rather than identify with, Hindi film protagonists usually read books of aspirational value: textbooks, English books, or high value literature."

Which one of the following CANNOT be inferred from the paragraph above?

(a) Protagonists in Hindi movies, being ideals for viewers, read only books of aspirational value
(b) People do not look up to writers of textbooks, English books or high value literature
(c) Though popular Hindi fiction has wide reach: it often does not appear in the movies
(d) Textbooks, English books or high literature have aspirational value, but not popular Hindi Fiction

[CE, 2019 : 2 Marks, Set-II]

193. The Newspaper reports that over 500 hectares of tribal land spread across 28 tribal settlements in Mohinitampuram forest division have already been "alienated". A top forest official said, "First the tribals are duped out of their land holdings. Second, the families thus rendered landless are often forced to encroach further into the forests".

On the basis of the information available in the paragraph, ________ is/are responsible for duping the tribals.

(a) The Newspaper
(b) it cannot be inferred who
(c) forest officials
(d) landless families

[CE, 2019 : 2 Marks, Set-II]

194. The expenditure on the project ________ as follows; equipments Rs. 20 lakhs, salaries Rs. 12 lakhs, and contingency Rs.3 lakhs.

(a) break
(b) breaks
(c) breaks down
(d) break down

[CS, 2019 : 1 Mark]

195. A court is to a judge as ________ is to a teacher.

(a) a syllabus
(b) a school
(c) a student
(d) a punishment

[CS, 2019 : 1 Mark]

196. The search engine's business model ________ around the fulcrum of trust.

(a) plays
(b) bursts
(c) revolves
(d) sinks

[CS, 2019 : 1 Mark]

197. "A recent High Court judgement has sought to dispel the idea of begging as a disease — which leads to its stigmatization and criminalization — and to regard it as a symptom. The underlying disease is the failure of the state to protect citizens who fall through the social security net".

Which one of the following statements can be inferred from the given passage?

(a) Begging is an offence that has to be dealt with firmly
(b) Beggars are created because of the lack of social welfare schemes
(c) Begging has to be banned because it adversely affects the welfare of the state
(d) Beggars are lazy people who beg because they are unwilling to work

[CS, 2019 : 2 Marks]

198. When he did not come home, she ______ him lying dead on the roadside somewhere.

(a) notice (b) looked
(c) concluded (d) pictured

[EC, 2019 : 1 Mark]

199. The strategies that the company ____ to sell its products _____ house-to house marketing.

(a) used, includes (b) uses, include
(c) use, includes (d) uses, including

[EC, 2019 : 1 Mark]

200. The boat arrived ______ dawn.

(a) on (b) at
(c) in (d) under

[EC, 2019 : 1 Mark]

201. "Indian history was written by British historians-extremely well documented and researched, but not always impartial. History had to serve its purpose: Everything was made subservient to the glory of the Union Jack. Latter-day Indian scholars presented a contrary picture."

From the text above, we can infer that:

Indian history written by British historians ______.

(a) was well documented and researched but was sometimes biased
(b) was well documented and not researched but was always biased
(c) was not well documented and researched and was sometimes biased
(d) was not well documented and researched and was always biased

[EC, 2019 : 2 Marks]

202. Newspapers are a constant source of delight and recreation for me. The ______ trouble is that I read ______ many of them.

(a) even, quite (b) only, too
(c) even, too (d) only, quite

[EE, 2019 : 1 Mark]

203. The passengers were angry ______ the airline staff about the delay.

(a) towards (b) on
(c) about (d) with

[EE, 2019 : 1 Mark]

204. I am not sure if the bus that has been booked will be able to ______ all the students.

(a) fill (b) sit
(c) accommodate (d) deteriorate

[EE, 2019 : 1 Mark]

205. An award-winning study by a group of researchers suggests that men are as prone to buying on impulse as women but women feel more guilty about shopping.

Which of the following statements can be inferred from the given text?

(a) Many men and women indulge in buying on impulse.
(b) Few men and women indulge in buying on impulse.
(c) All men and women indulge in buying on impulse.
(d) Some men and women indulge in buying on impulse.

[EE, 2019 : 2 Marks]

206. Consider five people - Mita, Ganga, Rekha, Lakshmi and Sana. Ganga is taller than both Rekha and Lakshmi. Lakshmi is taller than Sana. Mita is taller than Ganga. Which of the following conclusions are true?

1. Lakshmi is taller than Rekha
2. Rekha is shorter than Mita
3. Rekha is taller than Sana
4. Sana is shorter than Ganga

(a) 1 and 3 (b) 3 only
(c) 1 only (d) 2 and 4

[EE, 2019 : 2 Marks]

207. The fishermen, ______ the flood victims owed their lives, were rewarded by the government.

(a) whom (b) to which
(c) to whom (d) that

[IN, 2019 : 1 Mark]

208. Until Iran came along, India had never been________in kabaddi
(a) defeated (b) defeating
(c) defeat (d) defeatist
[IN, 2019 : 1 Mark]

209. Since the last one year, after a 125 basis point reduction in repo rate by the Reserve Bank of India, banking institutions have been making a demand to reduce interest rates on small saving schemes. Finally, the government announced yesterday a reduction in interest rates on small saving schemes to bring them on par with fixed deposit interest rates.

Which one of the following statements can be inferred from the given passage?
(a) Whenever the Reserve Bank of India reduces the repo rate, the interest rates on small saving schemes are also reduced
(b) Interest rates on small saving schemes are always maintained on par with fixed deposit interest rates
(c) The government sometimes takes into consideration the demands of banking institutions before reducing the interest rates on small saving schemes
(d) A reduction in interest rates on small saving schemes follow only after a reduction in repo rate by the Reserve Bank of India
[IN, 2019 : 2 Mark]

210. "I read somewhere that in ancient times the prestige of a kingdom depended upon the number of taxes that it was able to levy on its people. It was very much like the prestige of a head-hunter in his own community."

Based on the paragraph above, the prestige of a head-hunter depended upon_____
(a) the prestige of the kingdom
(b) the prestige of the heads
(c) the number of taxes he could levy
(d) the number of heads he could gather
[IN, 2019 : 2 Mark]

211. The nomenclature of Hindustani music has changed over the centuries. Since the medieval period *dhrupad* styles were identified as *baanis*. Terms like *gayaki* and *baaj* were used to refer to vocal and instrumental styles, respectively. With the institutionalization of music education the term *gharana* became acceptable. *Gharana* originally referred to hereditary musicians from a particular lineage, including disciples and grand disciples.

Which one of the following pairings is NOT correct?
(a) *dhrupad*, *baani* (b) *gayaki*, vocal
(c) *baaj*, institution (d) *gharana*, lineage
[IN, 2019 : 2 Mark]

212. John Thomas, an ______writer, passed away in 2018.
(a) prominent (b) dominant
(c) imminent (d) eminent
[ME, 2019 : 1 Mark, Set-1]

213. _________I permitted him to leave, I wouldn't have had any problem with him being absent, _______I?
(a) Had, wouldn't (b) Have, would
(c) Have, wouldn't (d) Had, would
[ME, 2019 : 1 Mark, Set-1]

214. The minister avoided any mention of the issue of women's reservation in the private sector. He was accused of ________ the issue.
(a) tying (b) collaring
(c) skirting (d) belting
[ME, 2019 : 1 Mark, Set-1]

215. Congo was named by Europeans. Congo's dictator Mobuto later changed the name of the country and the river to Zaire with the objective of Africanising names of persons and spaces. However, the name Zaire was a Portuguese alternation of *Nzadi o Nzere*, a local African term meaning 'River that swallows Rivers'. Zaire was the Portuguese name for the Congo river in the 16th and 17th centuries.

Which one of the following statements can be inferred from the paragraph above?
(a) The term *Nzadi o Nzere* was of Portuguese origin
(b) As a dictator Mobuto ordered the Portuguese to alter the name of the river to Zaire
(c) Mobuto's desire to Africanise names was prevented by the Portuguese
(d) Mobuto was not entirely successful in Africanising the name of his country
[ME, 2019 : 2 Mark, Set-1]

216. Under a certain legal system, prisoners are allowed to make one statement. If their statement turns out to be true then they are hanged. If the statement turns out to be false then they are shot. One prisoner made a statement and the judge had no option but to set him free. Which one of the following could be that statement?
(a) I committed the crime
(b) You committed the crime
(c) I will be shot
(d) I did not commit the crime

[ME, 2019 : 2 Mark, Set-1]

217. Are there enough seats here? There are ________ people here than I expected.
(a) most (b) more
(c) least (d) many

[ME, 2019 : 1 Mark, Set-2]

218. Once the team of analysts identify the problem, we ________ in a better position to comment on the issue.
Which one of the following choices CANNOT fill the given blank?
(a) were to be (b) are going to be
(c) will be (d) might be

[ME, 2019 : 1 Mark, Set-2]

219. A final examination is the _______ of a series of evaluations that a student has to go through.
(a) insinuation (b) desperation
(c) culmination (d) consultation

[ME, 2019 : 1 Mark, Set-2]

220. X is an online media provider. By offering unlimited and exclusive online content at attractive prices for a loyalty membership, X is almost forcing its customers towards its loyalty membership. If its loyalty membership continues to grow at its current rate, within the next eight years more households will be watching X than cable television.
Which one of the following statements can be inferred from the above paragraph?
(a) Most households that subscribe to X's loyalty membership discontinue watching cable television
(b) Cable television operators don't subscribe to X's loyalty membership
(c) The X is cancelling accounts of non-members
(d) Non-members prefer to watch cable television

[ME, 2019 : 2 Mark, Set-2]

221. While teaching a creative writing class in India. I was surprised at receiving stories from the students that were all set in distant places: in the American West with cowboys and in Manhattan penthouses with clinking ice cubes. This was, till an eminent Caribbean writer gave the once-colonized countries the confidence to see the shabby lives around them as worthy of being "told".
The writer of this passage is surprised by the creative writing assignments of his students, because ________
(a) None of the students had written about ice cubes and cowboys.
(b) None of the students had written stories set in India.
(c) Some of the students had written about ice cubes and cowboys.
(d) Some of the students had written stories set in foreign places.

[ME, 2019 : 2 Mark, Set-2]

222. The American psychologist Howard Gardner expounds that human intelligence can be sub-categorised into multiple kinds, in such a way that individuals differ with respect to their relative competence in each kind. Based on this theory, modern educationists insist on prescribing multi-dimensional curriculum and evaluation parameters that enable development and assessment of multiple intelligences.
Which of the following statements can be inferred from the given text?
(a) Modern educationists want to develop and asses the theory of multiple intelligences.
(b) Modern educationists insist that the teaching curriculum and evaluation needs to be multi-dimensional.
(c) Howard Gardner wants to develop and assess the theory of multiple intelligences.
(d) Howard Gardner insist that the teaching curriculum and evaluation needs to be multi-dimensional.

[CE, 2020 : 2 Marks, Set-1]

223. It is a common criticism that most of the academicians live in their ____, so, they are not aware of the real life challenges.
(a) glass palaces
(b) big flats
(c) ivory towers
(d) homes

[CE, 2020 : 1 Mark, Set-1]

224. Select the work that fits the analogy:
Fuse : Fusion :: Use : ____
(a) Usage (b) Uses
(c) Usion (d) User

[CE, 2020 : 1 Mark, Set-1]

225. His number for reading is insatiable. He reads indiscriminately. He is most certainly a/an ______ reader.
(a) all-round (b) voracious
(c) precocious (d) wise

[CE, 2020 : 1 Mark, Set-1]

226. Select the word that fits the analogy:
Partial : Impartial : : Popular : ____
(a) Impopular (b) Dispopular
(c) Mispopular (d) Unpopular

[CE, 2020 : 1 Mark, Set-2]

227. Select the most appropriate word that can replace the underlined word without changing the meaning of the sentence:
Now-a-days, most children have a tendency to <u>belittle</u> the legitimate concerns of their parents.
(a) Applaud (b) Begrudge
(c) Disparage (d) Reduce

[CE, 2020 : 1 Mark, Set-2]

228. Rescue teams deployed ______ disaster hit areas combat ______ a lot of difficulties to save the people.
(a) with, at (b) to, to
(c) with, with (d) in, with

[CE, 2020 : 1 Mark, Set-2]

229. Nominal interest rate is defined as the amount paid by the borrower to the lender for using the borrowed amount for a specific period of time. Real interest rate calculated on the basis of actual value (inflation-adjusted), is approximately equal to the difference between nominal rate and expected rate of inflation in the economy.
Which of the following assertions is best supported by the above information?
(a) Under low inflation, real interest rate is low and borrowers get benefited.
(b) Under high inflation, real interest rate is low and borrowers get benefited.
(c) Under low inflation, real interest rate is high and borrowers get benefited.
(d) Under high inflation, real interest rate is low and lenders get benefited.

[CE, 2020 : 2 Marks, Set-2]

230. He is known for his unscrupulous ways. He always sheds ______ tears to deceive people.
(a) crocodile's
(b) fox's
(c) crocodile
(d) fox

[IN & ME, 2020 : 1 Mark, Set-1]

231. Crowd funding deals with mobilisation of funds for a project from a large number of people, who would be willing to invest smaller amounts through web-based platforms in the project.
Based on the above paragraph, which of the following is correct about crowd funding?
(a) Funds raised through unwilling contributions on web-based platforms.
(b) Funds raised through coerced contributions on web-based platforms.
(c) Funds raised through large contributions on web-based platforms.
(d) Funds raised through voluntary contributions on web-based platforms.

[IN & ME, 2020 : 1 Mark, Set-1]

232. Jofra Archer, the England fast bowler, is ______ than accurate.
(a) faster (b) more faster
(c) more fast (d) less fast

[IN & ME, 2020 : 1 Mark, Set-1]

233. Select the word that fits the analogy:
Build : Building : : Grow : ______
(a) Grew (b) Growed
(c) Grown (d) Growth

[IN & ME, 2020 : 1 Mark, Set-1]

234. I do not think you know the case well enough to have opinions. Having said that, I agree with your other point. What does the phrase "having said that" mean in the given text?

(a) in addition to what I have said
(b) despite what I have said
(c) contrary to what I have said
(d) as opposed to what I have said

[IN & ME, 2020 : 1 Mark, Set-1]

235. The recent measures to improve the output would _____ the level of production to our satisfaction.

(a) increase (b) equalise
(c) decrease (d) speed

[PI & ME, 2020 : 1 Mark, Set-2]

236. While I agree ________ his proposal this time, I do not often agree ______ him.

(a) with, to (b) to, with
(c) with, with (d) to, to

[PI & ME, 2020 : 1 Mark, Set-2]

237. Select the word that fits the analogy:

White : Whitening : : Light : ________.

(a) Lightening (b) Lightning
(c) Enlightening (d) Lighting

[PI & ME, 2020 : 1 Mark, Set-2]

238. In one of the greatest innings ever seen in 142 years of Test history. Ben Stokes upped the tempo in a five-and-a-half hour long stay of 219 balls including 11 fours and 8 sixes that saw him finish on a 135 not out as England squared the five-match series.

Based on their connotations in the given passage, which one of the following meanings DOES NOT match?

(a) squared = lost
(b) tempo = enthusiasm
(c) upped = increased
(d) saw = resulted in

[PI & ME, 2020 : 1 Mark, Set-2]

239. Climate change and resilience deal with two aspects - reduction of sources of non-renewable energy resources and reducing vulnerability of climate change aspects. The terms 'mitigation' and 'adaptation' are used to refer to these aspects, respectively.

Which of the following assertions is best supported by the above information?

(a) Mitigation deals with actions taken to reduce the use of fossil fuels.
(c) Adaptation deals with causes of climate change.
(b) Mitigation deals with consequences of climate change.
(d) Adaptation deals with actions taken to combat green-house gas emissions

[PI & ME, 2020 : 1 Mark, Set-2]

240. Goods and Services Tax (GST) is an indirect tax introduced in India in 2017 that is imposed on the supply of goods and services, and it subsumes all indirect taxes except few. It is a destination-based tax imposed on goods and services used, and it is not imposed at the point of origin from where goods come. GST also has a few components specific to state governments, central government and Union Territories (UTs).

Which one of the following statements can be inferred from the given passage?

(a) GST is imposed on the production of goods and services.
(b) GST does not have a component specific to UT.
(c) GST is imposed at the point of usage of goods and services.
(d) GST includes all indirect taxes.

[CS, 2020 : 2 Marks]

241. Raman is confident of speaking English ________ six months as he has been practising regularly the last three weeks.

(a) for, since (b) within, for
(c) during, for (d) for, in

[CS, 2020 : 1 Mark]

242. His knowledge of the subject was excellent but his classroom performance was ________.

(a) extremely poor (b) praiseworthy
(c) desirable (d) good

[CS, 2020 : 1 Mark]

243. Select the word that fits the analogy:

Cook : Cook :: Fly : ________

(a) Flying (b) Flyer
(c) Flew (d) Flighter

[CS, 2020 : 1 Marks]

244. The dawn of the 21st century witnessed the melting glaciers oscillating between giving too much and too little to billions of people who depend on their for fresh water. The UN climate report estimates that without deep cuts to man-made emissions, at least 30% of the northern hemisphere's surface permafrost could melt by the end of the century. Given this situation of imminent global exodus of billions of people displaced by rising seas, nation-states need to rethink their carbon footprint for political concerns, if not for environmental ones.

Which one of the following statements can be inferred from the given passage?

(a) Billions of people are affected by melting glaciers.

(b) Billions of people are responsible for man-made emissions.

(c) Nation-states do not have environmental concerns.

(d) Nation-states are responsible for providing fresh water to billions of people.

[CS, 2020 : 1 Marks]

245. The Canadian constitution requires that equal importance be given to English and French. Last year, Air Canada lost a lawsuit, and had to pay a six-figure fine to a French-speaking couple after they filed complaints about formal in-flight announcements in English lasting 15 seconds, as opposed to informal 5 second messages in French.

The French-speaking couple were upset at ________.

(a) equal importance being given to English and French.

(b) the in-flight announcements being made in English.

(c) the English announcements being longer than the French ones.

(d) the English announcements being clearer than the French ones.

[EC, 2020 : 1 Marks]

246. Select the word that fits the analogy:

Explicit : Implicit : : Express : ________

(a) Repress (b) Suppress

(c) Compress (d) Impress

[EC, 2020 : 1 Mark]

247. The untimely loss of life is a cause of serious global concern as thousands of people get killed ______ accidents every year while many other die ______ diseases like cardio vascular disease, cancer, etc.

(a) during, from (b) in, of

(c) from, from (d) from, of

[EC, 2020 : 1 Mark]

248. The global financial crisis in 2008 is considered to be the most serious world-wide financial crisis, which started with the sub-prime lending crisis in USA in 2007. The sub-prime lending crisis led to the banking crisis in 2008 with the collapse of Lehman Brothers in 2008. The sub-prime lending refers to the provision of loans to those borrowers who may have difficulties in repaying loans, and it arises because of excess liquidity following the East Asian crisis.

Which one of the following sequences shows the correct precedence as per the given passage?

(a) Subprime lending crisis → global financial crisis → banking crisis → East Asian crisis.

(b) Banking crisis → subprime lending crisis → global financial crisis → East Asian crisis.

(c) East Asian crisis → subprime lending crisis → banking crisis → global financial crisis.

(d) Global financial crisis → East Asian crisis → banking crisis → subprime lending crisis.

[EC, 2020 : 2 Marks]

249. He was not only accused of theft ________ of conspiracy.

(a) rather (b) rather than

(c) but also (d) but even

[EC, 2020 : 1 Mark]

250. Stock markets ________ at the news of the coup.

(a) plunged (b) plugged

(c) poised (d) probed

[EE, 2020 : 1 Mark]

251. People were prohibited ________ their vehicles near the entrance of the main administrative building.

(a) to park (b) to have parked

(c) from parking (d) parking

[EE, 2020 : 1 Mark]

252. Select the word that fits the analogy:

Do : Undo : Trust : ______

(a) Untrust (b) Intrust
(c) Distrust (d) Entrust

[EE, 2020 : 1 Mark]

253. This book, including all its chapters ______ interesting. The students as well as the instructor ______ in agreement about it.

(a) are, are (b) is, was
(c) is, are (d) were, was

[EE, 2020 : 1 Mark]

254. Computers are ubiquitous. They are used to improve efficiency in almost all fields from agriculture to space exploration. Artificial intelligence (AI) is currently a hot topic. AI enables computers to learn, given enough training data. For humans, sitting in front of a computer for long hours can lead to health issues.

Which of the following can be deduced from the above passage?

(i) Nowadays, computers are present in almost all places.
(ii) Computers cannot be used for solving problems in engineering.
(iii) For humans, there are both positive and negative effects of using computers.
(iv) Artificial intelligence can be done without data.

(a) (ii) and (iii) (b) (i), (iii) and (iv)
(c) (i) and (iii) (d) (ii) and (iv)

[EC & PI, 2021 : 2 Marks]

255. Nostalgia is to anticipation as ______ is to ______.

Which one of the following options maintains a similar logical relation in the above sentence?

(a) Past, future (b) Future, past
(c) Future, present (d) Present, past

[EC & PI, 2021 : 1 Mark]

256. Consider the following sentences :

(i) I woke up from sleep.
(ii) I woked up from sleep.
(iii) I was woken up from sleep.
(iv) I was wokened up from sleep.

Which of the above sentences are grammatically CORRECT?

(a) (i) and (ii) (b) (ii) and (iii)
(c) (i) and (iv) (d) (i) and (iii)

[EC & PI, 2021 : 1 Mark]

257. Consider the following sentences:

(i) Everybody in the class is prepared for the exam.
(ii) Babu invited Danish to his home because he enjoys playing chess,

Which of the following is the CORRECT observation about the above two sentences?

(a) (i) is grammatically incorrect and (ii) is unambiguous
(b) (i) is grammatically correct and (ii) is ambiguous
(c) (i) is grammatically incorrect and (ii) is ambiguous
(d) (i) is grammatically correct and (ii) is unambiguous

[CS (Set-1), 2021 : 1 Mark]

258. Some people suggest anti-obesity measures (AOM) such as displaying calorie information in restaurant menus. Such measures sidestep addressing the core problems that cause obesity: poverty and income inequality.

Which one of the following statements summarizes the passage?

(a) The proposed AOM addresses the care problems that cause obesity.
(b) If obesity reduces, poverty will naturally reduce, since obesity causes poverty.
(c) AOM are addressing the problem superficially.
(d) AOM are addressing the core problems and are likely to succeed.

[CS (Set-1), 2021 : 2 Marks]

259. ________ is to surgery as writer is to ________

Which one of the following options maintains a similar logical relation in the above sentence?

(a) Plan, outline (b) Hospital, library
(c) Doctor, book (d) Medicine, grammar

[CS (Set-1), 2021 : 1 Mark]

260. Listening to music during exercise improves exercise performance and reduces discomfort. Scientists researched whether listening to music while studying can help students learn better and the results were inconclusive. Students who needed external stimulation for studying fared worse while students who did not need any external stimulation benefited from music.

Which one of the following statements is the CORRECT inference of the above passage?

(a) Listening to music has a clear positive effect on learning in all students. Music has a positive effect only in some students who exercise.
(b) Listening to music has a clear positive effect on physical exercise. Music has a positive effect on learning only in some students.
(c) Listening to music has a clear positive effect both on physical exercise and on learning.
(d) Listening to music has no effect on learning and a positive effect on physical exercise.

[CS (Set-2), 2021 : 2 Marks]

261. Pen : Write : : Knife : ______

Which one of the following options maintains a similar logical relation in the above?

(a) Sharp
(b) Cut
(c) Vegetables
(d) Blunt

[CS (Set-2), 2021 : 1 Mark]

262. Gauri said that she can play keyboard ______ her sister.

(a) as worse as
(b) as better as
(c) as well as
(d) as nicest as

[CS (Set-2), 2021 : 1 Mark]

263. Consider the following sentences:

1. After his surgery, Raja hardly could walk.
2. After his surgery, Raja could barely walk.
3. After his surgery, Raja barely could walk.
4. After his surgery, Raja could hardly walk.

Which of the above sentences are grammatically CORRECT?

(a) 1 and 2 (b) 1 and 3
(c) 2 and 4 (d) 3 and 4

[ME (Set-1), 2021 : 1 Mark]

264. Oxpeckers and rhinos manifest a symbiotic relationship in the wild. The oxpeckers warn the rhinos about approaching poachers, thus possibly saving the lives of the rhinos. Oxpeckers also feed on the parasitic ticks found on rhinos.

In the symbiotic relationship described above, the primary benefits for oxpeckers and rhinos respectively are,

(a) Oxpeckers get a food source, rhinos have no benefit.
(b) Oxpeckers save their habitat from poachers while the rhinos have no benefit.
(c) Oxpeckers get a food source, rhinos may be saved from the poachers.
(d) Oxpeckers save the lives of poachers, rhinos save their own lives.

[ME (Set-1), 2021 : 1 Mark]

265. "The increased consumption of leafy vegetables in the recent months is a clear indication that the people in the state have begun to lead a healthy lifestyle".

Which of the following can be logically inferred from the information presented in the above statement?

(a) Consumption of leafy vegetables may not be the only indicator of healthy lifestyle.
(b) The people in the state have increased awareness of health hazards causing by consumption of junk foods.
(c) Leading a healthy lifestyle is related to a diet with leafy vegetables.
(d) The people in the state did not consume leafy vegetables earlier.

[ME (Set-1), 2021 : 1 Mark]

266. The world is going through the worst pandemic in the past hundred years. The air travel industry is facing a crisis, as the resulting quarantine requirement for travellers led to weak demand.

In relation to the first sentence above, what does the second sentence do?

(a) Second sentence entirely contradicts the first sentence.
(b) States an effect of the first sentence.
(c) The two statements are unrelated.
(d) Restates an idea from the first sentence.

[ME (Set-2), 2021 : 2 Marks]

267. Consider the following sentences:

(i) The number of candidates who appear for the GATE examination is staggering.
(ii) A number of candidates from any class are appearing for the GATE examination.
(iii) The number of candidates who appear for the GATE examination are stagging.
(iv) A number of candidates from my class is appearing for the GATE examination.

Which of the above sentences are grammatically CORRECT?

(a) (i) and (iii)
(b) (i) and (ii)
(c) (ii) and (iii)
(d) (ii) and (iv)

[ME (Set-2), 2021 : 1 Mark]

268. The people ______ were at the demonstration were from all section of society.

(a) whose (b) who
(c) whom (d) which

[EE, 2021 : 1 Mark]

269. The importance of sleep is often overlooked by students when they are preparing for exams. Research has consistently shown that sleep deprivation greatly reduces the ability to recall the material learn. Hence, cutting down on sleep to study longer hours can be counterproductive.

Which one of the following statements is the CORRECT inference from the above passage?

(a) Sleeping well alone is enough to prepare for an exam. Studying has lesser benefit.
(b) To do well in an exam, adequate sleep must be part of the preparation.
(c) Students are efficient and are not wrong in thinking that sleep is a waste of time.
(d) If a student is extremely well prepared for an exam, he needs little or no sleep.

[EE, 2021 : 2 Marks]

270. Oasis is to sand as island is to ______ .

Which one of the following options maintains a similar logical relation in the above sentence?

(a) Mountain (b) Water
(c) Land (d) Stone

[EE, 2021 : 1 Mark]

271. Human have the ability to construct worlds entirely in their minds, which don't exist in the physical world. So far as we know, no other species possesses this ability. This skill is so important that we have different worlds to refer to its different flavors, such as imagination invention and innovation.

Based on the above passage, which one of the following is TRUE?

(a) We do not know of any species other than humans who posses the ability to construct mental worlds.
(b) Imagination, invention and innovation are unrelated to the ability to construct mental worlds.
(c) No spices posses the ability to construct worlds in their minds.
(d) The terms imagination, invention and innovation refer to unrelated skill.

[CE (Set-1) & IN, 2021 : 1 Mark]

272. Getting to the top is ______ than staying on top.

(a) more easy (b) easier
(c) much easy (d) easiest

[CE (Set-1) & IN, 2021 : 1 Mark]

273. The author said, "Musicians rehearse before their concerts. Actors rehearse their roles before the opening of a new play. On the other hand, I find it strange that many public speakers think they can just walk on to the stage and start speaking. In my opinion, it is no less important for public speakers to rehearse their talks."

Based on the above passage, which one of the following is TRUE?

(a) The author is of the opinion that rehearsing is less important for public speakers than for musicians and actors.
(b) The author is of the opinion that rehearsing is more important only for musicians than public speakers.
(c) The author is of the opinion that rehearsal is more important for actors than musicians.
(d) The author is of the opinion that rehearsing is important for musicians, actors and public speakers.

[CE (Set-2), 2021 : 2 Mark]

274. (i) Arun and Aparna are here.
(ii) Arun and Aparna is here.
(iii) Arun's families is here.
(iv) Arun's family is here.

Which of the above sentences are grammatically CORRECT?

(a) (i) and (iv) (b) (iii) and (iv)
(c) (i) and (ii) (d) (ii) and (iv)

[CE (Set-2), 2021 : 1 Mark]

275. You should ______ when to say ______.

(a) no/no (b) no/know
(c) know/know (d) know/no

[CE : (Set-I), 2022 : 1 Mark]

276 Healthy eating is a critical component of healthy aging. When should one start eating healthy? It turns out that it is never too early. For example, babies who start eating healthy in the first year are more likely to have better overall health as they get older.

Which one of the following is the CORRECT logical inference based on the information in the above passage?

(a) Healthy eating is important for those with good health conditions, but not for others
(b) Eating healthy can be started at any age, earlier the better
(c) Eating healthy and better overall health are more correlated at a young age, but not older age
(d) Healthy eating is more important for adults than kids

[CE (Set-I), 2022 : 2 Marks]

277 The movie was funny and I _______ .

(a) could helped laughed
(b) couldn't help laughed
(c) couldn't help laughing
(d) could help laughing

[CE (Set-II), 2022 : 1 Mark]

278 In the last few years, several new shopping malls were opened in the city. The total number of visitors in the malls is impressive. However, the total revenue generated through sales in the shops in these malls is generally low.

Which one of the following is the CORRECT logical inference based on the information in the above passage?

(a) More people are visiting the malls and spending more
(b) Fewer people are visiting the malls but spending more
(c) More people are visiting the malls but not spending enough
(d) Fewer people are visiting the malls and not spending enough

[CE : (Set-II), 2022 : 2 Marks]

279. After playing _______ hours of tennis, I am feeling _______ tired to walk back.

(a) too/too (b) too/two
(c) two/two (d) two/too

[ME : (Set-I), 2022 : 1 Mark]

280. Writing too many things on the _____ while teaching could make students get _____.

(a) bored/board (b) board/bored
(c) board/board (d) bored/bored

[ME : (Set-II), 2022 : 1 Mark]

281. Humans are naturally compassionate and honest. In a study using strategically placed wallets that appear "lost", it was found that wallets with money are more likely to be returned than wallets without money. Similarly, wallets that had a key and money are more likely to be returned than wallets with the same amount of money alone. This suggests that the primary reason for this behavior is compassion and empathy.

Which one of the following is the CORRECT logical inference based on the information in the above passage?

(a) Wallets with a key are more likely to be returned because people do not care about money.
(b) Wallets with a key are more likely to be returned because people relate to suffering of others.
(c) Wallets used in experiments are more likely to be returned than wallets that are really lost.
(d) Money is always more important than keys.

[ME : (Set-I), 2022 : 2 Marks]

282. Fish belonging to species S in the deep sea have skins that are extremely black (ultra-black skin). This helps them not only to avoid predators but also sneakily attack their prey. However, having this extra layer of black pigment results in lower collagen on their skin, making their skin more fragile.

Which one of the following is the CORRECT logical inference based on the information in the above passage?

(a) Having ultra-black skin is only advantageous to species S.
(b) Species S with lower collagen in their skin are at an advantage because it helps them avoid predators.
(c) Having ultra-black skin has both advantages and disadvantages to species S.
(d) Having ultra-black skin is only disadvantageous to species S but advantageous only to their predators.

[ME : (Set-II), 2022 : 2 Marks]

283. Mr. X speaks ________ Japanese ________ Chinese.

(a) neither / or (b) either / nor
(c) neither / nor (d) also / but

[EC, 2022 : 1 Mark]

284. Mosquitoes pose a threat to human health. Controlling mosquitoes using chemicals may have undesired consequences. In Florida, authorities have used genetically modified mosquitoes to control the overall mosquito population. It remains to be seen if this nobel approach has unforeseen consequences.

Which one of the following is the correct logical inference based on the information in the above passage?

(a) Using chemicals to kill mosquitoes is better than using genetically modified mosquitoes because genetic engineering is dangerous.
(b) Using genetically modified mosquitoes is better than using chemicals to kill mosquitoes because they do not have any side effects.
(c) Both using genetically modified mosquitoes and chemicals have undesired consequences and can be dangerous.
(d) Using chemicals to kill mosquitoes may have undesired consequences but it is not clear if using genetically modified mosquitoes has any negative consequence.

[EC, 2022 : 2 Marks]

285. As you grow older, an injury to your _____ may take longer to _____.

(a) heel / heel (b) heal / heel
(c) heal / heal (d) heel / heal

[EE, 2022 : 1 Mark]

286. The ________ is too high for it to be considered ________.

(a) fair / fare (b) faer / fair
(c) fare / fare (d) fare / fair

[CS, 2022 : 1 Mark]

287 Some people believe that "what gets measured, improves". Some others believe that "what gets measured, gets gamed". One possible reason for the difference in the beliefs is the work culture in organizations. In organizations with good work culture, metrics help improve outcomes. However, the same metrics are counterproductive in organizations with poor work culture.

Which one of the following is the CORRECT logical inference based on the information in the above passage?

(a) Metrics are useful in organizations with poor work culture.
(b) Metrics are useful in organizations with good work culture.
(c) Metrics are always counterproductive in organizations with good work culture.
(d) Metrics are never useful in organizations with good work culture.

[CS, 2022 : 2 Marks]

288. Inhaling the smoke from a burning ________ could ________ you quickly.

(a) tire / tier (b) tire / tyre
(c) tyre / tire (d) tyre / tier

[CH, IN & PI, 2022 : 1 Mark]

289. Rice, a versatile and inexpensive source of carbohydrate, is a critical component of diet worldwide. Climate change, causing extreme weather, poses a threat to sustained availability of rice. Scientists are working on developing Green Super Rice (GSR), which is resilient under extreme weather conditions yet gives higher yields sustainably.

Which one of the following is the CORRECT logical inference based on the information given in the above passage?

(a) GSR is an alternative to regular rice, but it grows only in an extreme weather.
(b) GSR may be used in future in response to adverse effects of climate change.
(c) GSR grows in an extreme weather, but the quantity of produce is lesser than regular rice.
(d) Regular rice will continue to provide good yields even in extreme weather.

[CH, IN & PI, 2022 : 2 Marks]

290. Rafi told Mary, "I am thinking of watching a film this weekend."

The following reports the above statement in indirect speech :

Rafi told Mary that he ____ of watching a film that weedkend.

(a) thought (b) is thinking
(c) am thinking (d) was thinking

[EE, 2023 : 1 Mark]

291. Permit : _____ :: Enforce : Relax (by word meaning)

(a) Allow (b) Forbid
(c) License (d) Reinforce

[EE, 2023 : 1 Mark]

292. Which one of the options does NOT describe the passage below or follow from it?

We tend to think of cancer as a 'modern' illness because its metaphors are so modern. It is a disease of overproduction, of sudden growth, a growth that is unstoppable, tipped into the abyss of no control. Modern cell biology encourages us to imagine the cell as a molecular machine. Cancer is that machine unable to quench its initial command (to grow) and thus transform into an indestructible, self-propelled automation.

(Adapted from The Emperor of All Maladies by Siddhartha Mukherjee)

(a) It is a reflection of why cancer seems so modern to most of us.
(b) It tells us that modern cell biology uses and promotes metaphors of machinery.
(c) Modern cell biology encourages metaphors of machinery, and cancer is often imagined as a machine.
(d) Modern cell biology never uses figurative language, such as metaphors, to describe or explain anything.

[EE, 2023 : 2 Marks]

293. A recent survey shows that 65% of tobacco users were advised to stop consuming tobacco. The survey also shows that 3 out of 10 tobacco users attempted to stop using tobacco.

Based only on the information in the above passage, which one of the following options can be logically inferred with certainty?

(a) A majority of tobacco users who were advised to stop consuming tobacco made an attempt to do so.
(b) A majority of tobacco users who were advised to stop consuming tobacco did not attempt to do so.
(c) Approximately 30% of tobacco users successfully stopped consuming tobacco.
(d) Approximately 65% of tobacco users successfully stopped consuming tobacco.

[EE, 2023 : 1 Mark]

294. Students of all the departments of a college who have successfully completed the registration process are eligible to vote in the upcoming college elections. However, by the time the due date for registration was over, it was found that surprisingly none of the students from the Department of Human Sciences had completed the registration process.

Based only on the information provided above, which one of the following sets of statement(s) can be logically inferred with certainty?

(i) All those students who would not be eligible to vote in the college elections would certainly belong to the Department of Human Sciences.
(ii) None of the students from departments other than Human Sciences failed to complete the registration process within the due time.
(iii) All the eligible voters would certainly be students who are not from the Department of Human Sciences.

(a) (i) and (ii) (b) (i) and (iii)
(c) Only (i) (d) Only (iii)

[EE, 2023 : 2 Marks]

295. He did not manage to fix the car himself, so he ____ in the garage.

(a) got it fixed (b) getting it fixed
(c) gets fixed (d) got fixed

[ME, 2023 : 1 Mark]

296. Planting : Seed : : Raising : ______. [By word meaning]

(a) Child (b) Temperature
(c) Height (d) Lift

[ME, 2023 : 1 Mark]

297. In a recently held parent-teacher meeting, the teachers had very few complaints about Ravi. After all, Ravi was a hardworking and kind student. Incidentally, almost all of Ravi's friends at school were hardworking and kind too. But the teachers drew attention to Ravi's complete lack of interest in sports. The teachers believed that, along with some of his friends who showed similar disinterest in sports, Ravi needed to engage in some sports for his overall development.

Based only on the information provided above, which one of the following statements can be logically inferred with certainty?

(a) All of Ravi's friends are hardworking and kind.
(b) No one who is not a friend of Ravi is hardworking and kind.
(c) None of Ravi's friends are interested in sports.
(d) Some of Ravi's friends are hardworking and kind.

[ME, 2023 : 2 Marks]

298. Which one of the sentence sequences in the given options creates a coherent narrative?
(i) I could not bring myself to knock.
(ii) There was a murmur of unfamiliar voices coming from the big drawing room and the door was firmly shut.
(iii) The passage was dark for a bit, but then it suddenly opened into a bright kitchen.
(iv) I decided I would rather wander down the passage.
(a) (iv), (i), (iii), (ii) (b) (iii), (i), (ii), (iv)
(c) (ii), (i), (iv), (iii) (d) (i), (iii), (ii), (iv)

[ME, 2023 : 2 Marks]

299. The village was nestled in a green spot, _______ the ocean and the hills.
(a) through (b) in
(c) at (d) between

[IN, 2023 : 1 Mark]

300. Disagree : Protest : : Agree : _______.
(By word meaning)
(a) Refuse (b) Pretext
(c) Recommend (d) Refute

[IN, 2023 : 1 Mark]

301. Residency is a famous housing complex with many well-established individuals among its residents. A recent survey conducted among the residents of the complex revealed that all of those residents who are well established in their respective fields happen to be academicians. The survey also revealed that most of these academicians are authors of some best-selling books.

Based only on the information provided above, which one of the following statements can be logically inferred with certainty?
(a) Some residents of the complex who are well established in their fields are also authors of some best-selling books.
(b) All academicians residing in the complex are well established in their fields.
(c) Some authors of best-selling books are residents of the complex who are well established in their fields.
(d) Some academicians residing in the complex are well established in their fields.

[IN, 2023 : 2 Marks]

302. The information contained in DNA is used to synthesize proteins that are necessary for the functioning of life. DNA is composed of four nucleotides: Adenine (A), Thymine (T), Cytosine (C), and Guanine (G). The information contained in DNA can then be thought of as a sequence of these four nucleotides: A, T, C, and G. DNA has coding and non-coding regions. Coding regions—where the sequence of these nucleotides are read in groups of three to produce individual amino acids—constitute only about 2% of human DNA. For example, the triplet of nucleotides CCG codes for the amino acid glycine, while the triplet GGA codes for the amino acid proline. Multiple amino acids are then assembled to form a protein.

Based only on the information provided above, which of the following statements can be logically inferred with certainty?
(i) The majority of human DNA has no role in the synthesis of proteins.
(ii) The function of about 98% of human DNA is not understood.
(a) only (i) (b) only (ii)
(c) both (i) and (ii) (d) neither (i) nor (ii)

[IN, 2023 : 2 Marks]

303. "I cannot support this proposal. My _______ will not permit it."
(a) conscious (b) consensus
(c) conscience (d) consent

[EC, 2023 : 1 Mark]

304 Courts : _______ : : Parliament : Legislature
(a) Judiciary (b) Executive
(c) Governmental (d) Legal

[EC, 2023 : 1 Mark]

305. Which one of the following options can be inferred from the given passage alone?

When I was a kid, I was partial to stories about other worlds and interplanetary travel. I used to imagine that I could just gaze off into space and be whisked to another planet.

[Excerpt from The Truth about Stories by T. King]

(a) It is a child's description of what he or she likes.
(b) It is an adult's memory of what he or she liked as a child.
(c) The child in the passage read stories about interplanetary travel only in parts.
(d) It teaches us that stories are good for children.

[EC, 2023 : 2 Marks]

306. Out of 1000 individuals in a town, 100 unidentified individuals are covid positive. Due to lack of adequate covid-testing kits, the health authorities of the town devised a strategy to identify these covid-positive individuals. The strategy is to:

(i) Collect saliva samples from all 1000 individuals and randomly group them into sets of 5.
(ii) Mix the samples within each set and test the mixed sample for covid.
(iii) If the test done in (ii) gives a negative result, then declare all the 5 individuals to be covid negative.
(iv) If the test done in (ii) gives a positive result, then all the 5 individuals are separately tested for covid.

Given this strategy, no more than _______ testing kits will be required to identify all the 100 covid positive individuals irrespective of how they are grouped.

(a) 700 (b) 600
(c) 800 (d) 1000

[EC, 2023 : 2 Marks]

307. Forestland is a planet inhabited by different kinds of creatures. Among other creatures, it is populated by animals all of whom are ferocious. There are also creatures that have claws, and some that do not. All creatures that have claws are ferocious.

Based only on the information provided above, which one of the following options can be logically inferred with certainty?

(a) All creatures with claws are animals.
(b) Some creatures with claws are non-ferocious.
(c) Some non-ferocious creatures have claws.
(d) Some ferocious creatures are creatures with claws.

[EC, 2023 : 2 Marks]

308. We reached the station late, and _______ missed the train.

(a) near (b) nearly
(c) utterly (d) mostly

[CS, 2023 : 1 Mark]

309. Kind : _______ : : Often : Frequently
(By word meaning)

(a) Mean (b) Type
(c) Cruel (d) Kindly

[CS, 2023 : 1 Mark]

310. Which one of the following sentence sequences creates a coherent narrative?

(i) Once on the terrace, on her way to her small room in the corner, she notices the man right away.
(ii) She begins to pant by the time she has climbed all the stairs.
(iii) Mina has bought vegetables and rice at the market, so her bags are heavy.
(iv) He was leaning against the parapet, watching the traffic below.

(a) (i), (ii), (iv), (iii) (b) (ii), (iii), (i), (iv)
(c) (iv), (ii), (i), (iii) (d) (iii), (ii), (i), (iv)

[CS, 2023 : 2 Marks]

311. A survey for a certain year found that 90% of pregnant women received medical care at least once before giving birth. Of these women, 60% received medical care from doctors, while 40% received medical care from other healthcare providers.

Given this information, which one of the following statements can be inferred with certainty?

(a) More than half of the pregnant women received medical care at least once from a doctor.
(b) Less than half of the pregnant women received medical care at least once from a doctor.
(c) More than half of the pregnant women received medical care at most once from a doctor.
(d) Less than half of the pregnant women received medical care at most once from a doctor.

[CS, 2023 : 1 Mark]

312. The country of Zombieland is in distress since more than 75% of its working population is suffering from serious health issues. Studies conducted by competent health experts concluded that a complete lack of physical exercise among its working population was one of the leading causes of their health issues. As one of the measures to address the problem, the Government of Zombieland has decided to provide monetary incentives to those who ride bicycles to work.

Based only on the information provided above, which one of the following statements can be logically inferred with certainty?

(a) All the working population of Zombieland will henceforth ride bicycles to work.
(b) Riding bicycles will ensure that all of the working population of Zombieland is free of health issues.
(c) The health experts suggested to the Government of Zombieland to declare riding bicycles as mandatory.
(d) The Government of Zombieland believes that riding bicycles is a form of physical exercise.

[CS, 2023 : 2 Marks]

313. "You are delaying the completion of the task. Send _______ contributions at the earliest."

(a) you are (b) your
(c) you're (d) yore

[CH + PI 2023 : 1 Mark]

314. References : ______ : : Guidelines : Implement (By word meaning)

(a) Sight (b) Site
(c) Cite (d) Plagiarise

[CH + PI 2023 : 1 Mark]

315. In 2022, June Huh was awarded the Fields medal, which is the highest prize in Mathematics.

When he was younger, he was also a poet. He did not win any medals in the International Mathematics Olympiads. He dropped out of college.

Based only on the above information, which one of the following statements can be logically inferred with certainty?

(a) Every Fields medalist has won a medal in an International Mathematics Olympiad.
(b) Everyone who has dropped out of college has won the Fields medal.
(c) All Fields medalists are part-time poets.
(d) Some Fields medalists have dropped out of college.

[CH + PI 2023 : 1 Mark]

316. Human beings are one among many creatures that inhabit an imagined world. In this imagined world, some creatures are cruel. If in this imagined world, it is given that the statement "Some human beings are not cruel creatures" is FALSE, then which of the following set of statement(s) can be logically inferred with certainty?

(i) All human beings are cruel creatures.
(ii) Some human beings are cruel creatures.
(iii) Some creatures that are cruel are human beings.
(iv) No human beings are cruel creatures.

(a) only (i) (b) only (iii) and (iv)
(c) only (i) and (ii) (d) (i), (ii) and (iii)

[CH + PI 2023 : 2 Marks]

317. The World Bank has declared that it does not plan to offer new financing to Sri Lanka, which is battling its worst economic crisis in decades, until the country has an adequate macroeconomic policy framework in place. In a statement, the World Bank said Sri Lanka needed to adopt structural reforms that focus on economic stabilisation and tackle the root causes of its crisis. The latter has starved it of foreign exchange and led to shortages of food, fuel, and medicines. The bank is repurposing resources under existing loans to help alleviate shortages of essential items such as medicine, cooking gas, fertiliser, meals for children, and cash for vulnerable households.

Based only on the above passage, which one of the following statements can be inferred with certainty?

(a) According to the World Bank, the root cause of Sri Lanka's economic crisis is that it does not have enough foreign exchange.
(b) The World Bank has stated that it will advise the Sri Lankan government about how to tackle the root causes of its economic crisis.

(c) According to the World Bank, Sri Lanka does not yet have an adequate macroeconomic policy framework.
(d) The World Bank has stated that it will provide Sri Lanka with additional funds for essentials such as food, fuel, and medicines.

[CH + PI 2023 : 2 Marks]

318. "I have not yet decided what I will do this evening; I ______ visit a friend."

(a) mite (b) would
(c) might (d) didn't

[CE, 2023 : 1 Mark, Forenoon]

319. Eject : Insert : : Advance : ______ (By word meaning)

(a) Advent (b) Progress
(c) Retreat (d) Loan

[CE, 2023 : 1 Mark, Forenoon]

320. A duck named Donald Duck says "All ducks always lie."

Based only on the information above, which one of the following statements can be logically inferred with certainty?

(a) Donald Duck always lies.
(b) Donald Duck always tells the truth.
(c) Donald Duck's statement is true.
(d) Donald Duck's statement is false.

[CE, 2023 : 1 Mark, Forenoon]

321. The James Webb telescope, recently launched in space, is giving humankind unprecedented access to the depths of time by imaging very old stars formed almost 13 billion years ago. Astrophysicists and cosmologists believe that this odyssey in space may even shed light on the existence of dark matter. Dark matter is supposed to interact only via the gravitational interaction and not through the electromagnetic-, the weak- or the strong-interaction. This may justify the epithet "dark" in dark matter.

Based on the above paragraph, which one of the following statements is FALSE?

(a) No other telescope has captured images of stars older than those captured by the James Webb telescope.
(b) People other than astrophysicists and cosmologists may also believe in the existence of dark matter.
(c) The James Webb telescope could be of use in the research on dark matter.
(d) If dark matter was known to interact via the strong-interaction, then the epithet "dark" would be justified.

[CE, 2023 : 2 Marks, Forenoon]

322. The line ran ______ the page, right through the centre, and divided the page into two.

(a) across (b) of
(c) between (d) about

[CE, 2023 : 1 Mark, Afternoon]

323. Kind : ______ : : Often : Seldom (By word meaning)

(a) Cruel (b) Variety
(c) Type (d) Kindred

[CE, 2023 : 1 Mark, Afternoon]

324. Elvesland is a country that has peculiar beliefs and practices. They express almost all their emotions by gifting flowers. For instance, if anyone gifts a white flower to someone, then it is always taken to be a declaration of one's love for that person. In a similar manner, the gifting of a yellow flower to someone often means that one is angry with that person.

Based only on the information provided above, which one of the following sets of statement(s) can be logically inferred with certainty?

(i) In Elvesland, one always declares one's love by gifting a white flower.
(ii) In Elvesland, all emotions are declared by gifting flowers.
(iii) In Elvesland, sometimes one expresses one's anger by gifting a flower that is not yellow.
(iv) In Elvesland, sometimes one expresses one's love by gifting a white flower.

(a) only (ii) (b) (i), (ii) and (iii)
(c) (i), (iii) and (iv) (d) only (iv)

[CE, 2023 : 2 Marks, Afternoon]

325. Based only on the following passage, which one of the options can be inferred with certainty?

When the congregation sang together, Apenyo would also join, though her little screams were not quite audible because of the group singing. But whenever there was a special number, trouble would begin; Apenyo would try singing along, much to the embarrassment of her mother. After two or three such mortifying Sunday evenings, the mother stopped going to church altogether until Apenyo became older and learnt to behave.

At home too, Apenyo never kept quiet; she hummed or made up silly songs to sing by herself, which annoyed her mother at times but most often made her become pensive. She was by now convinced that her daughter had inherited her love of singing from her father who had died unexpectedly away from home.

[Excerpt from These Hills Called Home by Temsula Ao]

(a) The mother was embarrassed about her daughter's singing at home.
(b) The mother's feelings about her daughter's singing at home were only of annoyance.
(c) The mother was not sure if Apenyo had inherited her love of singing from her father.
(d) When Apenyo hummed at home, her mother tended to become thoughtful.

[CE, 2023 : 2 Marks, Afternoon]

326. If '→' denotes increasing order of intensity, then the meaning of the words [sick → infirm → moribund] is analogous to [silly → ______ → daft].

Which one of the given options is appropriate to fill the blank?

(a) frown (b) fawn
(c) vein (d) vain

[PI, GATE-2024 : 1 Mark]

327. Thousands of years ago, some people began dairy farming. This coincided with a number of mutations in a particular gene that resulted in these people developing the ability to digest dairy milk.

Based on the given passage, which of the following can be inferred?

(a) All human beings can digest dairy milk.
(b) No human being can digest dairy milk.
(c) Digestion of dairy milk is essential for human beings.
(d) In human beings, digestion of dairy milk resulted from a mutated gene.

[PI, GATE-2024 : 2 Marks]

328. In the given text, the blanks are numbered (i)-(iv). Select the best match for all the blanks.

Prof. P ___(i)___ merely a man who narrated funny stories ___(ii)___ in his blackest moments he was capable of self-depreciating humor.

Prof. Q ___(iii)___ a man who hardly narrated funny stories ___(iv)___ in his blackest moments was he able to find humor.

(a) (i) wasn't (ii) Even, (iii) was, (iv) Only
(b) (i) wasn't (ii) Only, (iii) was, (iv) Even
(c) (i) was (ii) Only, (iii) wasn't, (iv) Even
(d) (i) was (ii) Even, (iii) wasn't, (iv) Only

[ME, GATE-2024 : 1 Mark]

329. If '→' denotes increasing order of intensity, then the meaning of the words

[smile → giggle → laugh] is analogous to [disapprove → ______ → chide].

Which one of the given options is appropriate to fill the blank?

(a) reprove (b) grieve
(c) reprise (d) praise

[ME, GATE-2024 : 2 Marks]

330. In the given text, the blanks are numbered (i)-(iv). Select the best match for all the blanks.

From the ancient Athenian arena to the modern Olympic stadiums, athletics ___(i)___ the potential for a spectacle. The crowd ___(ii)___ with bated breath as the Olympian artist twists his body, stretching the javelin behind him. Twelve strides in, he begins to cross-step. Six cross-steps ___(iii)___ in an abrupt stop on his left foot. As his body ___(iv)___ like a door turning on a hinge, the javelin is launched skyward at a precise angle.

(a) (i) holds (ii) wait (iii) culminates (iv) pivot
(b) (i) hold (ii) wait (iii) culminate (iv) pivots
(c) (i) holds (ii) waits (iii) culminate (iv) pivots
(d) (i) hold (ii) Waits (iii) culminates (iv) pivot

[CE1 + CH, GATE-2024 : 2 Marks]

331. If ' → ' denotes increasing order of intensity, then the meaning of the words [simmer → seethe → smolder] is analogous to [break → raze → ______]. Which one of the given options is appropriate to fill the blank?

(a) obliterate (b) fissure
(c) fracture (d) obfuscate

[CE1 + CH, GATE-2024 : 1 Mark]

332. If '→' denotes increasing order of intensity, then the meaning of the words [drizzle → rain → downpour] is analogous to [______ → quarrel → feud].

Which one of the given options is appropriate to fill the blank?

(a) bicker (b) bog
(c) dither (d) dodge

[IN + CE2, GATE-2024 : 1 Mark]

333. In the given text, the blanks are numbered (i)-(iv). Select the best match for all the blanks.

Yoko Roi stands ___(i)___ as an author for standing ___(ii)___ as an honorary fellow, after she stood ___(iii)___ her writings that stand ___(iv)___ the freedom of speech.

(a) (i) out (ii) down (iii) in (iv) for
(b) (i) down (ii) out (iii) by (iv) in
(c) (i) down (ii) out (iii) for (iv) in
(d) (i) out (ii) down (iii) by (iv) for

[IN + CE2, GATE-2024 : 2 Marks]

334. In the given text, the blanks are numbered (i) - (iv). Select the best match for all the blanks.

Steve was advised to keep his head ___(i)___ before heading ___(ii)___ to bat; for, while he had a head ___(iii)___ batting, he could only do so with a cool head ___(iv)___ his shoulders.

(a) (i) on (ii) down (iii) for (iv) on
(b) (i) down (ii) down (iii) on (iv) for
(c) (i) down (ii) out (iii) for (iv) on
(d) (i) on (ii) out (iii) on (iv) for

[CS, GATE-2024 : 2 Marks, Forenoon]

335. If '→' denotes increasing order of intensity, then the meaning of the words

[dry → arid → parched] is analogous to [diet → fast → ______].

Which one of the given options is appropriate to fill the blank?

(a) reject (b) feast
(c) deny (d) starve

[CS, GATE-2024 : 1 Mark, Forenoon]

336. If '→' denotes increasing order of intensity, then the meaning of the words [walk → jog → sprint] is analogous to [bothered → ______ → daunted].

Which one of the given options is appropriate to fill the blank?

(a) phased (b) phrased
(c) fazed (d) fused

[CS, GATE-2024 : 1 Mark, Afternoon]

337. Sequence the following sentences (P, Q, R, S) in a coherent passage:

P: Shifu's student exclaimed, "Why do you run since the bull is an illusion?"
Q: Shifu said. "Surely my running away from the bull is also an illusion."
R: Shifu once proclaimed that all life is illusion.
S: One day, when a bull gave him chase. Shifu began running for his life.

(a) RSPQ (b) SPRQ
(c) SRPQ (d) RPQS

[EC, GATE-2024 : 2 Marks]

338. If '→' denotes increasing order of intensity, then the meaning of the words [charm → enamor → bewitch] is analogous to [bored → ____ → weary].

Which one of the given options is appropriate to fill the blank?

(a) worsted (b) dead
(c) jaded (d) baffled

[EC, GATE-2024 : 1 Mark]

339. If '→' denotes increasing order of intensity, then the meaning of the words [talk → shout → scream] is analogous to [please → ______ → pander].

Which one of the given options is appropriate to fill the blank?

(a) flutter (b) fritter
(c) frizzle (d) flatter

[EE, GATE-2024 : 1 Mark]

340. In the given text, the blanks are numbered (i)-(iv). Select the best match for all the blanks.

Following a row ___(i)___ the shopkeeper ___(ii)___ the price of a frying pan, the cook stood ___(iii)___ a row to withdraw cash ___(iv)___ the ATM booth.

(a) (i) with; (ii) over; (iii) in; (iv) at
(b) (i) with; (ii) over; (iii) at; (iv) with
(c) (i) over; (ii) with; (iii) over; (iv) at
(d) (i) at; (ii) over; (iii) over; (iv) in

[EE, GATE-2024 : 2 Marks]

■■■■

Answer Key — **GATE :** Previous Years Solved Questions

1. (b)
2. (a)
3. (d)
4. (c)
5. (d)
6. (b)
7. (a)
8. (c)
9. (b)
10. (c)
11. (d)
12. (c)
13. (b)
14. (d)
15. (b)
16. (b)
17. (a)
18. (d)
19. (a)
20. (d)
21. (a)
22. (a)
23. (b)
24. (d)
25. (a)
26. (b)
27. (c)
28. (c)
29. (b)
30. (d)
31. (a)
32. (a)
33. (b)
34. (c)
35. (d)
36. (d)
37. (c)
38. (b)
39. (d)
40. (d)
41. (c)
42. (d)
43. (b)
44. (b)
45. (b)
46. (c)
47. (c)
48. (b)
49. (b)
50. (a)
51. (b)
52. (b)
53. (b)
54. (c)
55. (c)
56. (c)
57. (b)
58. (a)
59. (b)
60. (a)
61. (b)
62. (d)
63. (d)
64. (c)
65. (c)
66. (b)
67. (d)
68. (a)
69. (b)
70. (d)
71. (d)
72. (a)
73. (c)
74. (c)
75. (b)
76. (b)
77. (c)
78. (b)
79. (d)
80. (b)
81. (b)
82. (d)
83. (c)
84. (d)
85. (c)
86. (a)
87. (b)
88. (a)
89. (c)
90. (c)
91. (a)
92. (a)
93. (c)
94. (b)
95. (a)
96. (b)
97. (b)
98. (a)
99. (b)
100. (c)
101. (b)
102. (b)
103. (b)
104. (b)
105. (a)
106. (b)
107. (a)
108. (b)
109. (c)
110. (c)
111. (a)
112. (b)
113. (a)
114. (b)
115. (c)
116. (b)
117. (c)
118. (a)
119. (b)
120. (c)
121. (c)
122. (a)
123. (d)
124. (a)
125. (a)
126. (b)
127. (b)
128. (a)
129. (c)
130. (d)
131. (d)
132. (c)
133. (a)
134. (c)
135. (d)
136. (c)
137. (b)
138. (b)
139. (b)
140. (a)
141. (b)
142. (a)
143. (b)
144. (c)
145. (c)
146. (c)
147. (b)
148. (b)
149. (b)
150. (a)
151. (c)
152. (d)
153. (b)
154. (b)
155. (d)
156. (c)
157. (a)
158. (a)
159. (d)
160. (a)
161. (c)
162. (c)
163. (d)
164. (d)
165. (b)
166. (c)
167. (a)
168. (a)
169. (a)
170. (a)
171. (a)
172. (b)
173. (a)
174. (b)
175. (a)
176. (a)
177. (d)
178. (c)
179. (a)
180. (c)
181. (a)
182. (d)
183. (a)
184. (a)
185. (d)
186. (c)
187. (c)
188. (a)
189. (d)
190. (d)
191. (c)
192. (b)
193. (b)
194. (c)
195. (b)
196. (c)
197. (b)
198. (d)
199. (b)
200. (b)
201. (a)
202. (b)
203. (d)
204. (c)
205. (d)
206. (d)
207. (c)
208. (a)
209. (c)
210. (d)
211. (c)
212. (d)
213. (d)
214. (c)
215. (d)
216. (c)
217. (b)
218. (a)
219. (c)
220. (a)
221. (b)
222. (b)
223. (c)
224. (a)
225. (b)
226. (d)
227. (c)
228. (d)
229. (b)
230. (c)
231. (d)
232. (c)
233. (d)
234. (b)
235. (a)
236. (b)
237. (a)
238. (a)
239. (a)
240. (c)
241. (b)
242. (a)
243. (b)
244. (a)
245. (c)
246. (a)
247. (b)
248. (c)
249. (c)
250. (a)
251. (c)
252. (c)

Answer Key — GATE : Previous Years Solved Questions

253. (c)	268. (b)	283. (c)	298. (c)	313. (b)	328. (a)
254. (c)	269. (b)	284. (d)	299. (d)	314. (c)	329. (a)
255. (a)	270. (b)	285. (d)	300. (c)	315. (d)	330. (c)
256. (d)	271. (a)	286. (d)	301. (*)	316. (d)	331. (a)
257. (b)	272. (b)	287. (b)	302. (d)	317. (c)	332. (a)
258. (c)	273. (d)	288. (c)	303. (c)	318. (c)	333. (d)
259. (c)	274. (a)	289. (b)	304. (a)	319. (c)	334. (c)
260. (b)	275. (d)	290. (d)	305. (b)	320. (d)	335. (d)
261. (b)	276. (b)	291. (b)	306. (a)	321. (d)	336. (c)
262. (c)	277. (c)	292. (d)	307. (d)	322. (a)	337. (a)
263. (c)	278. (c)	293. (b)	308. (b)	323. (a)	338. (c)
264. (c)	279. (d)	294. (d)	309. (b, d)	324. (d)	339. (d)
265. (c)	280. (b)	295. (a)	310. (d)	325. (d)	340. (a)
266. (b)	281. (b)	296. (a)	311. (a)	326. (d)	
267. (b)	282. (c)	297. (d)	312. (d)	327. (d)	

Explanations

1. (b)
Circuitous: Deviating from a straight course ⇒ Indirect
(a) Cyclic: Recurring in cycle
(b) Indirect: Not leading by straight line
(c) Confusing: Lacking clarity
(d) Crooked: For shapes (irregular in shape)

2. (a)
Unemployed: Worker ⇒ Here one is opposite to other.
(a) Fallow: Land ⇒ Fallow means undeveloped land.
(b) Unaware: sleeper ⇒ Both are same unaware or asleep.
(c) Wit: Jester ⇒ Wit means ability to make jokes and jester is a joker.
(d) Renovated : House ⇒ Renovate means to make better and house can be renovated.

3. (d)
(a) Uphold: to support ⇒ not appropriate
(b) Restrain: keep under control ⇒ not appropriate
(c) Cherish: be fond of ⇒ not related
(d) Conserve: Keep in safety and protect from harm, decay, loss, or destruction ⇒ most appropriate.

4. (c)
(a) Masked: Hide under a false appearance ⇒ opposite
(b) Belied: Be in contradiction with ⇒ not appropriate
(c) Betrayed: Reveal unintentionally ⇒ most appropriate
(d) Suppressed: To put down by force or authority ⇒ irrelevant

5. (d)
(a) Modern warfare has resulted in civil strife: There is no direct consequence of warfare given, so it is not appropriate.
(b) Chemical agents are useful in modern warfare: Passage does not say whether chemical agents are useful or not, so not appropriate.
(c) Use of chemical agents in warfare would be undesirable: Given that people in military think these are useful, undesirable is wrong
(d) People in military establishments like to use chemical agents in war; Correct choice as last statement tells that military people think that chemical agents are useful tools for their cause (work silently in warfare).

16. (b)
Setbacks: Despite several setbacks the mission succeeded in its attempt to resolve the conflict.
The word 'Despite' indicates that there has to be a contrast in the sentence, use of the word 'Setbacks' in the blank indicates that despite many problems the mission was successful.

17. (a)
Diminish: Mitigate means to reduce, to lessen etc. So only the word Diminish is close. Rest all choices have no link with the given word.
Divulge means to disclose or reveal which has no link with the given word.

19. (a)
Suresh's dog is the one that was hurt in the stampede. That is used with restrictive clauses.

20. (d)
Not gender-discriminatory: Choice (a) cannot be considered since there is no gender discrimination mentioned in the argument.
Choice (b) Xenophobic is one who has fear of foreigners, no link with the given argument.
Choice (c) It is wrong to say that the profile has not been designed to make the post attractive, since there are certain features which have been added to make the profile lucrative (which are given towards the end of the advertisement, like Transportation paid, expenses reimbursed).

23. (b)
Latitude refers to freedom of action, freedom of expression from restrictions etc. For example, he allowed his children a fair amount of latitude. Coercion refers to force which is an opposite of the word Latitude. Meticulousness refers to being extremely careful and conscientious.

24. **(d)**
Given the seriousness of the situation that he had to face, his nonchalance was impressive.
Nonchalance refers to casual / non-serious/ indifferent attitude. The author wants to convey that in spite of the seriousness of the situation, he was very casual.
No other word fits in the blank. Beggary is word that is related to extreme poverty (Beggar). Nomenclature refers to the art of naming.

25. **(a)**
Thorough regimentation was the main reason for the efficiency of the Roman legions even in adverse circumstances. As the author says, 'Discipline on the battlefield kept units obedient, intact and fighting, even when the odds and conditions were against them.' So choice (a) emerges as the best answer option.
We have been asked to summarize the paragraph in the best possible manner. Choice (c) presents one-sided picture of the paragraph. Summary is the essence or the gist of the paragraph.
Choice (b) and Choice (d) are wrong as per the information mentioned in the paragraph.

26. **(b)**
Friendship, No matter how intimate it is, has its limitation. Intimate refers to close personal relations, ex. an intimate friend. It is also characterized by or involving warm friendship or a personally close or familiar association or feeling.

27. **(c)**
Medicine leads to good health. Similarly, Education leads to Knowledge. Science does not lead to experiment. Wealth may not necessarily lead to peace. Also money may not also lead to happiness all the time.

28. **(c)**
Primeval pertains to the first age or ages, especially of the world. Ex: primeval forms of life.

31. **(a)**
Specificity: Universalism and Particularism are opposite words. Similarly, we require a word that indicates an opposite to diffuseness. Diffuseness refers to spreading or causing to spread in all directions. Specificity is an appropriate opposite.

32. **(a)**
Would fly: This is a sentence based on Conditional Sentences. In the first clause imagination (Were) has been used, so the main clause should comprise of 'would + I form of verb' usage.

33. **(b)**
Lowest: Nadir which is opposite to the word Zenith (the highest point) means the lowest point.

34. **(c)**
She is an European: The usage of articles a and an is not decided by the first alphabet of the word but the sound / pronunciation of the word. European though starts with 'E' which is a vowel but it does not yield a vowel sound. When read aloud, it leaves the sound of 'Y' which is a consonant sound.
So the correct sentence should be, "She is a European".

35. **(d)**
No adversity justifies giving up hope.
Options (b) and (c) can be clearly discarded since honesty and adventure making have no relationship with the given argument.
Option (a) may appear close. But remember, every failure may not result into success as it happens in the case of spider.

36. **(d)**
This is a famous idiom in English.

37. **(c)**
'You can always give me a ring whenever you need' implies that I am available even at a single call (whenever you need me). 'Ring' has been used as a metaphor here and not literally.

40. **(d)**
As the given argument states that there were different streams of freedom movement during colonialism, it clearly implies that nationalism in India is heterogeneous.
Option (c) is ruled out as it states the contrary.
Options (a) and (b) fail to establish a direct link with the argument.

109. **(c)**
Fish out of water: implies uncomfortable situation.

110. (c)

Increase in productivity necessarily reflects greater efforts made, by the employees.

145. (c)

The bacteria is in milk are destroyed when it is heated to 80° Celsius. Use of simple present tense is apt for indication factual statements / scientifically proven statements / universal truth.

146. (c)

Tampering with someone else's email account is a very serious offence. Tampering refers to interfering in a harmful or disruptive manner.

148. (b)

Use of conditional sentence based on past perfect structure. Conditional clause uses (had + V3), main clause is based on would / should / could + have + V3.

149. (b)

Thorough refers to somethings that is exhaustively complete, accurate, careful or absolute.

150. (a)

People coming out in the same order in which they enter indicates that the centre operates on a first-come-first-serve basis.

151. (c)

Correct option is; Wished

After Rajender Chola **returned** from his voyage to Indonesia, he **wished** to visit the temple in Thanjavur.

Both are events of past. Use of past perfect form is unwarranted as it reflects part of past.

152. (d)

Besides money

Research in the work place reveals the people works for many reasons besides money.

Besides conveys the meaning of 'in addition'

Beside means 'next to'

153. (b)

History is viewed through the filter of rationalism.

154. (b)

Bright is an adjective that is to express a bold object.

For *eg.*: She is bright enough to know that this is a very good offer.

Similarly, the statement will be saturn is bright enough to be seen on a clear night with the naked eye.

155. (d)

- Yielding means "giving a product or generating a financial return of a specified amount."
- Resistant means "to be immune or imprevious to something."

By the meaning, it can be observed that both are not synonyms to each other.

156. (c)

'Culture' and 'Political' words are not related to the belief that ideology is not as important. Similarly generally means 'as far as one known' and is the correct phrase that has been used by the author.

157. (a)

Could have, been

I could have made arrangements had I been informed earlier. Use of conditional sentence based on past participle form.

158. (a)

Hurtful

Hurtful means causing pain or suffering or something that is damaging or harmful.

The expression 'sharp tongue' defines a bitter or critical manner of speaking.

159. (d)

Hostile

Antagonist is a adversary or one who opposes/ contends against another.

164. (d)

Heated means angry, vehement

Which is reflected by speakers becoming agitated.

165. (b)

Steve was known to be better player than Mark.

166. (c)

Subject – The ways
Verb – are

167. (a)

Forgo: Means to go without something/refrain from

Exclusive: Excluding or not admitting other things

168. (a)

Brake is a device which is used for stopping or moving a vehicle.

Break refers to a pause in work or during an activity.

169. (a)

Unpredictable - Contrary word required

170. (a)

Shone: It is past-participle and past form of shine.

Shown: To show means to reveal and point out something.

171. (a)

Innovative is similar to pioneer.

172. (b)

They're used for pointing group.

Their is pointing people.

There is used for place.

173. (a)

Meandering: wandering aimlessly/indirect.

174. (b)

- Peace is a situation when there is no war or fighting and things are calm and quiet.
- Piece is a part of something.

175. (a)

Even though there is a vast scope for its improvement, tourism has remained a neglected area.

176. (a)

Going off the course - not following the intended route.

Coarse sand - harsh in texture

178. (c)

Errant means misbehaving, exhibiting inappropriate behavior / offending conduct.

179. (a)

Stairs: A construction designed to bridge a large vertical distance by dividing it into smaller vertical distances, called steps

Stares: To look at someone for a long time

180. (c)

Frugality is synonyms to miserliness

181. (a)

Principle: A moral rule / belief

Principal: The person in-charge of an educational institution.

182. (d)

Complemented - Complete

Complimented - Remark of admiration

183. (a)

Undiminished - Unambitious / unskilled.

184. (a)

They have came long way in creating trust among the users.

185. (d)

The lecture was attended by quite a few students so the hall was not very quiet.

- quite a few indicates a fairly large number of units.
- quiet refer to making little or no noise.

186. (c)

CEO's decision to quit was a shocking to the board as it was me.

- Board and me are receivers of the action hence objective case of pronoun (me) is to be used.

187. (c)

Daytime temperatures in Delhi can **reach** 40°.

188. (a)

Hima Das was **the** only Indian athlete to win **a** gold for India.

189. (d)

The growth of ABC motors was same in 2017 **as that of** *XYZ* motors in 2016.

Subject of the sentence is 'The growth of *ABC* motors'. It is to be compared with the growth of *XYZ* motors hence use of 'that of' is appropriate.

190. (d)

The paragraph deals with the increasing interest in tribal characters and further goes on to describe that tribal characters haven't attained any special status. There have been very few stories woven around these characters.

191. (c)
Farmers want to access the new technology even if it is not legal.
The argument emphasizes on the widespread use of the new cotton technology despite it being illegal. Detail about price is just an additional detail.

193. (b)
The Newspaper is just reporting the matter, it cannot be responsible for duping the tribals.
A top forest official made statement about tribals being duped hence officials cannot be responsible for duping the tribals.

194. (c)
'Breaks down' is a transitive phrasal verb which means to divide something such as a total amount into separate parts.
Option (c) is most appropriate.

195. (b)
A 'court' is for a 'judge' as a 'school' is for a 'teacher'.
Court is a place where a judge works.
Similarly, school is a place where a teacher works.

196. (c)
The search engine business model <u>revolves</u> around the fulcrum of trust.
Fulcrum is any thing that plays a central or essential role in an activity, event, or situation. Thus, out of the given options, 'revolves' is the most appropriate.

197. (b)
The passage states that the underlying disease behind begging is the failure of the state to protect citizens who fall through the social security net.
Only option (b) can be inferred from this.

198. (d)
When he did not come home, she pictured him lying dead on the roadside somewhere. "pictured" is the most appropriate option in this situation as she is imagining or creating an image of somebody/something in her mind.

199. (b)
"Company" is singular, hence, "uses" would be the correct option in blank 1. "products" is plural, hence, "include" would be the correct option in blank 2.

200. (b)
Option 1 i.e. "on" is used when one is talking about a day of the week e. g. on Thursday, or a particular part of a date e.g. on Sunday evening, or a particular date e.g. on 9th March.
Option 3 i.e. "in" is used when one is talking about a month e.g. in June, or a season e.g. in winter, or a specific year e.g. in 2019.
Option 4 i.e. "under" is absurd in this situation.
Option 2 i.e. "at" is used when one is referring to a particular time e.g. at dawn, or a calendar festival season e.g. at Christmas.

201. (a)
The passage says" extremely well documented and researched, but not always impartial". This suggests that Indian history written by British historians was well documented and researched but was sometimes biased. hence option (a) is the correct answer.

203. (d)
Angry on ... an issue
angry with ... someone.

206. (d)
If '>' Implies taller than
1. Ganga > Rekha, Ganga > Lakshmi
2. Lakshmi > Sana
3. Mita > Ganga
   ⇒ Mita > Ganga > Lakshmi > Sana
   ⇒ Mita > Ganga > Rekha

Statement 2 is correct.
Statement 4 is correct.

208. (a)
Until Iran came along, India had never been defeated in Kabaddi.
If two events occur in past one after another, the event completing first takes past perfect tense. And another event is expressed in Simple past tense.

209. (c)
The argument says that banking institutions had been demanding for a reduction in interest rates for the last one year. Finally, the government decided to reduce the interest rates of small saving schemes thus implying that the govt. does consider the demands of banking institutions before making any such policy decision.

210. (d)
The way prestige of a kingdom depended upon the number of taxes, the prestige of a head-hunter depended upon the number of heads he could collect.

211. (c)
Baaj, institution
Following are the correct pairings:
- Dhrupad style - baani
- gayaki - vocal style
- baaj - instrumental style
- gharana - musicians from a particular lineage

212. (d)
Eminent means one who is famous and respected within a particular sphere; distinguished or renowned.

213. (d)
Had, I permitted him to leave, I wouldn't have had any problem
Past perfect conditional sentence pattern.

$\underbrace{\text{Had} + V_3 \text{ form,}}_{\text{If clause}}\ \underbrace{\text{would have} + V_3 \text{ form,}}_{\text{main clause}}$

Negative statement requires affirmative question tag so would I?

214. (c)
He was accused of skirting.
Skirt as a verb means to avoid discussing a subject or problem, usually because there are difficulties that you do not want to deal with.

215. (d)
As the paragraph states, dictator Mobuto wanted to Africanise the name of Congo but could not succeed as the new name given to the country was Portuguese alteration of some other term.

216. (c)
Remaining three options would lead to the prisoner being hanged or shot. This answer choice creates two contradictory situations.

217. (b)
Comparative degree adjective is required since comparison is indicated (use of 'than').

218. (a)
Since the 1st clause is in present tense, the second clause will not use past tense.
are going to be, will be, might be fill the given blank appropriately.

219. (c)
Culmination refers to the end point or the final result of a process.

220. (a)
Most households that subscribe to X's loyalty membership discontinue watching cable television. It is estimated that if X continues to offer loyalty membership, the number of subscribers of X will outgrow that of cable television. It is based on the assumption that most of the households subscribing to X don't watch cable television any longer.

221. (b)
None of the students had written stories set in India. It is explicitly stated that the writer was surprised at receiving stories from the students that were all set in distant places.

226. (d)
Partial and Impartial are opposite, in the same way popular and unpopular are opposite words.

227. (c)
Belittle means to undervalue/ underestimate some as unimportant. Disparage fits in the most appropriate manner.

230. (c)
The term crocodile tears refers to hypocrite crying. It is a false or insincere display of grief.

231. (d)
Crowd funding has been defined as a large number of people making small contribution for a project. Only option and is implied.

232. (c)
When two qualities of the same noun are compared, more + positive degree adjective is used. Use more fast and not 'faster'.

233. (d)

Build *(verb)*	:	Building *(noun)*
Grow *(verb)*	:	Growth *(noun)*

236. (b)
Agree with - a person,
Agree to - an idea, proposal

240. (c)
According to passage GST is imposed on the supply of goods and services hence option (a) is incorrect which states that GST is imposed on production of goods and services.

In the passage it is mentioned that GST has a two components specific to UTs. So option (b) is incorrect.

Passage states that GST is a destination based tax which means it is imposed at the point of usage of goods and services. Hence (c) is correct.

Option (d) says GST includes all indirect taxes but as per passage it subsumes all indirect taxes **except few**.

241. (b)
- 'within' is a preposition that is used to express something that occurs inside a particular period of time.
- 'for' is used here because
  (i) Sentence is in 'present perfect continuous tense'.
  (ii) For is used when we talk about a period of time.

242. (a)
'But' is used for introducing an idea which contrasts with the statement that has been already said.

243. (b)
(i) Relation is verb : noun
(ii) One who cooks is a cook, similarly one who flies any aircraft is a flyer.

252. (c)
Do : Undo : Trust : Distrust

254. (c)
Ubiquitous is the keyword to justify option (i). Positive and negative effect for humans justifies option (iii).

255. (a)
Nostalgia refers to a feeling, fondness and slight sadness thinking of past and anticipate is to predicting future.

256. (d)
Wake - Woke - Woken are three form of the verb. Hence, option (i) and (iii) are correct.

257. (b)
Everybody is singular and takes singular verb is, first statement is correct and the purpose of invitation is unclear so second statement is vague and that sounds ambiguous.

258. (c)
Superficially is the deciding key word which means apparently/seemingly.

259. (c)
Doctor performs surgery as writer writes book.

260. (b)
"Only in some students" is the key in option (b) and that matches well with the given informations in the passage.

261. (b)
Pen is used to write and knife is used to cut.

262. (c)
(as – as) and (so – as) are used in positive degree of comparison only. Worse, better are comparative degree and nicest is superlative degree, which are not fit between as–as and so–as.

263. (c)
Hardly/Scarcely/Barely have same sense that is negative and they are used after the verb (could barely and could hardly).

264. (c)
Option (a) and (b) are weekend by expression 'rhinos have no benefit'. Oxpeckers do not save life of poachers, so option (d) is incorrect. Hence, option (c) is correct.

265. (c)
The last sentence of the passage is reflecting in option (c) only.

266. (b)
First option is wrong because second sentence does not contradict the first sentence. Third option is wrong because two sentences are related. Fourth option is wrong because the second sentence does not repeat the first one. Hence second option is correct which shows the result of the cause.

267. (b)
"The number of" is singular and it takes singular verb. "A number of " is plural and it takes plural verb.

268. (b)

We cannot use 'which' for people, 'who' is the only word which acts as a subject for the verb.

269. (b)

First statement is ridiculous, third and fourth are contradictory to what is given in the passage.

270. (b)

Oasis is surrounded by sand as Island is surrounded by water.

271. (a)

Option (b) and (d) are weekend by the word 'unrelated skills'. Option (c) is weekend by the expression, no species posses the ability.

Hence answer is option (a) which reflects the information given in the passage.

272. (b)

When the comparison is between two things we use the second degree of the adjective.

The degree form of easy are: (easy - easier - easiest).

273. (d)

The last sentence of the passage decides the answer with the key words "No Less Important".

274. (a)

Two subject joined with 'and' become plural and hence plural verb is there in first statement, in fourth sentence the subject is family which is singular and takes singular verb.

277. (c)

The movies was funny and I couldn't help laughing.

i.e. movie was so funny that the person couldn't control his/her laugh.

280. (b)

Writing too many things on the board while teaching could make students get bored.

283. (c)

Mr. X speaks neither Japanese nor Chinese.

284. (d)

The given passage mentions that it is yet to be seen whether the genetically modified mosquitoes to control the overall mosquito population has any unforeseen consequences whereas controlling mosquitoes using chemicals may have undesired consequences.

Hence, option (d) is the correct answer.

289. (b)

GSR may be used in future in response to adverse effects of climate change.

291. (b)

Forbid is antonym of permit.

293. (b)

Tobacco users advised to stop consumption = 65%

Tobacco users attempted to stop consumption = 3/10 = 30%

Majority of users who were advised to stop consumption did not attempt to do so.

326. (d)

[Sick → Infirm → Moribund] is anologous to [silly → vain → daft]

329. (a)

[smile → giggle → laugh] is similarly related as [disapprove → reprove → chide].

332. (a)

Bicker → To argue about unimportant thing or issue

Quarrel → Angry, argument or disagreement

Feud → Angry or serious argument between two people that continues for long period of time.

337. (a)

- Sentence R gives the introduction to passage.
- Sentence Q is the response to a question in sentence P, which is asked on the statement made in sentence S.

Hence, the correct sequence is RSPQ.

338. (c)

Jaded means tired or bored with something, because of experiencing it too much.

■■■■